SpringerBriefs in Earth System Sciences

South America and the Southern Hemisphere

Series editors

Gerrit Lohmann, Bremen, Germany
Lawrence A. Mysak, Montreal, Canada
Justus Notholt, Bremen, Germany
Jorge Rabassa, Ushuaia, Argentina
Vikram Unnithan, Bremen, Germany

More information about this series at http://www.springer.com/series/10032

Martín Ubilla · Sergio Martínez

Geology and Paleontology of the Quaternary of Uruguay

 Springer

Martín Ubilla
Facultad de Ciencias, Departamento
 de Paleontología
Instituto de Ciencias Geológicas
Montevideo
Uruguay

Sergio Martínez
Facultad de Ciencias, Departamento
 de Paleontología
Instituto de Ciencias Geológicas
Montevideo
Uruguay

ISSN 2191-589X   ISSN 2191-5903   (electronic)
SpringerBriefs in Earth System Sciences
ISBN 978-3-319-29301-1   ISBN 978-3-319-29303-5   (eBook)
DOI 10.1007/978-3-319-29303-5

Library of Congress Control Number: 2016930284

Printed on acid-free paper

This Springer imprint is published by SpringerNature
The registered company is Springer International Publishing AG Switzerland

Preface

Despite the Quaternary beds of Uruguay have an important geographical expression and economical significance, efforts to increase knowledge about them are far from exhaustive, implying that many more studies must be carried out. The benefit in having none, or slightly modified beds preserving most of the features, is the opposite of the short interval of geological time to be analyzed. Multidisciplinary studies are needed, integrating such disciplines as geology, paleontology, and geomorphology among others, in order to properly interpret the biotic and landscape evolution during the Pleistocene and Holocene.

This present contribution provides an updated review and criticism of the progress on the continental, transitional, and marine Quaternary deposits of Uruguay, mainly based on the analysis of the geological and paleontological evidence. Time (numerical ages and biostratigraphy), environment, and the climatic situations involved will be synthesized, and the most relevant paleontological contents highlighted. Chapter 2 addresses the information related to marine and related transitional beds, while Chap. 3 deals with the continental strata and some transitional beds. Illustrations on the geographic location of outcrops, the most relevant sedimentological features, and significant fossiliferous contents are provided. The numerical ages are organized in tables showing sample type and reference labs. Finally, Chap. 4 is a general consideration of regional correlations of the Uruguayan beds, highlighting some methodological problems related to the topic.

We are indebted to Jorge Rabassa for asking us to write this contribution. Responsible of Uruguayan collections allowed us to study fossil materials on many occasions (L. Castiglioni, M. Ghizzoni, F. López, A. Rinderknecht, A. Rojas, J. Soloviy, A. Sanchez, and M. Trindade). A. Currant, curator of the Mammal Fossil section of the Natural History Museum in London, facilitated M. Ubílla's access to the holotype of *Toxodon platensis* included in Fig. 1.1. We are grateful to A. Rojas, M. Verde, and C. Goso for providing the photographs in Fig. 2.7, and to

U. Meneghin for valuable information. We also thank our many colleagues for field assistance and the discussions in this book (F. Cabrera, A. Corona, G. Daners, C. Goso, E. Morosi, D. Perea, A. Ribeiro, A. Rinderknecht, A. Rojas, M. Verde, and G. Veroslavsky). The research of the authors has greatly benefited from the financial support of CSIC (UDELAR), ANII and PEDECIBA.

Contents

Chapter 1
Introduction

Abstract Nineteenth-century naturalists coming on expeditions from the Northern Hemisphere provided the first widely known observations and studies of Quaternary strata and fossils of Uruguay. A. d´Orbigny and C. Darwin visited this land and collected geological and paleontological information that was published at a later date. However, D.A. Larrañaga was the first naturalist to write about geological and paleontological topics of this country (although he has largely been forgotten). The early-to mid-twentieth century was characterized by many publications because the government urged the study of Uruguay's natural resources. In those times, most authors referred to the Quaternary beds of Uruguay as "Pampeano," "Post-Pampeano," and "transgresión querandina," under the influence of the Argentinean F. Ameghino. It was in the late twentieth century that authors departed from this context. Then, many new terms to identify different Quaternary beds appeared, some of which are still utilized today.

Keywords Quaternary · Uruguay · d´Orbigny · Darwin · Pampeano · Post-Pampeano · *Transgresión querandina*

1.1 Pathfinders

Nineteenth-century naturalists coming on expeditions from the Northern Hemisphere provided the first widely known observations and studies of Quaternary strata and fossils of Uruguay. d´Orbigny (1842a, b) visited this part of South America (1826–1827) and collected material for the Musée d´Histoire Naturelle of Paris, which was later published in several volumes. Darwin (1839, 1846) also visited this country and collected various kinds of information, including paleontological material. In 1833 he obtained from the Uruguyan Río Negro Department an almost complete skull, utilized by Owen (1837) for describing the extinct giant mammal *Toxodon platensis* (Fig. 1.1). It seems that in the early nineteenth century, Sellow collected the first fossils from the Late Pleistocene beds of northern Uruguay (a giant turtle and some mammals probably from the Paysandú

© United Nations University 2016
M. Ubilla and S. Martínez, *Geology and Paleontology of the Quaternary of Uruguay*, SpringerBriefs in Earth System Sciences,
DOI 10.1007/978-3-319-29303-5_1

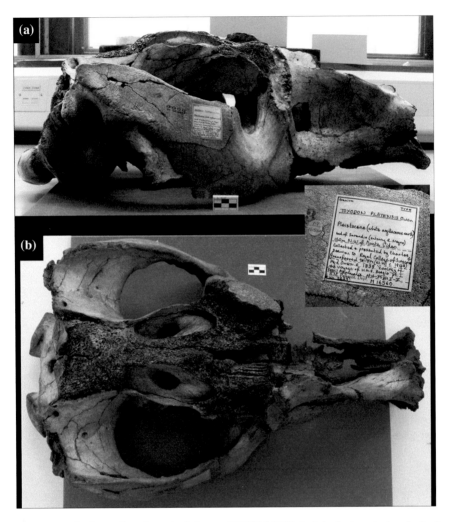

Fig. 1.1 Skull of *Toxodon platensis* (holotype BMNH-16560) (**a** lateral view, **b** dorsal view) housed at the Natural History Museum, London, obtained by Darwin in Uruguay and described by Owen (1837)

or Salto Departments) (Paula Couto 1948). However, the priest Larrañaga (1894) was probably the first naturalist to write about geological and paleontological topics from there, very likely around 1819; they included designs of glyptodonts and extinct proboscideans and recognized the displacement of the Rio de la Plata during the Quaternary transgressions (Méndez-Alzola 1950; Martínez and Ubilla 2004). Due to political instability of the region at those times, and characterized by independence campaigns, his observations were finally published in 1894.

In the early to mid-twentieth century, several studies with commentaries on the geological and paleontological Uruguayan Quaternary beds were published. A remarkable issue is that most of these publications included detailed and stratigraphic sections that are still useful today. In fact, these studies represent important efforts by the government to improve the knowledge of the natural resources of the country. A pioneer was the German Karl Walther, who developed an intense activity related to geology (Walther 1919, but see also the compilations of Mones 1981 and Veroslavsky et al. 2004). Sedimentary beds were mostly known as "Pampeano," "post-Pampeano" and "*transgresión querandina*" (Walther 1915; Kraglievich 1928, 1932; Frenguelli 1930; Teisseire 1930; Lambert 1941; Serra 1943, 1944, 1945, and Jones 1956, among others). In fact, these names were introduced by d'Orbigny ("*argile pampéenne*") and Darwin ("Pampean mud"), but the influence of the Argentinean Florentino Ameghino (1854–1911) was instrumental in the use of this terminology. The Pampean unit was substituted by Caorsi and Goñi (1958) by "*loess de Arazatí*" and Post-pampian as "*arcillas grises del Vizcaíno*." In the late twentieth century, the aforementioned conception was modified in various publications, and a large number of names were proposed and some are still utilized today (Bossi 1966; Bossi and Navarro 1991; Goso 1965, 1972; Goso and Bossi 1966; Preciozzi et al. 1985; among others) (see Chaps. 2 and 3) (Fig. 1.2).

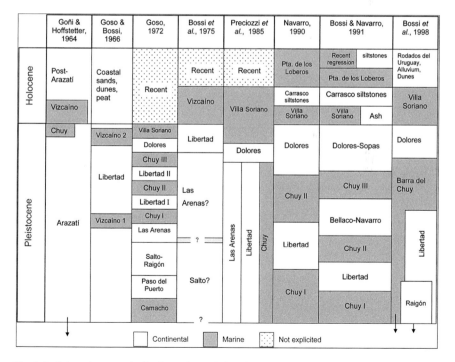

Fig. 1.2 Selected proposal of units and correlation for the continental and marine Quaternary of Uruguay (adapted from Martínez et al. 2001)

1.2   Nomenclature Problems

The Quaternary sedimentary beds of Uruguay have been classified according to a large number of names. Unfortunately, some lithostratigraphic units are seldom undefined and do not correctly use the formal requirements.

Chronological, geomorphological, or paleontological criteria have been used to distinguish lithological units, characterizing, in fact, local lithological facies and sometimes creating confusion. In addition, the identity and extension of some of the units differ according to each author's concept. Consequently, there is a panoply of names for the same beds (Fig. 1.2).

In this book, we include most of the common names used in the available literature and authors' research. It is not an exhaustive examination of all units described, but rather a selected study highlighting those with a more geographical view and fossil content.

References

Bossi J (1966) Geología del Uruguay. Dept Public Univ Rep, Uruguay

Bossi J, Navarro R (1991) Geología del Uruguay. Dept Publ Univ Rep. 2, Montevideo

Caorsi JH, Goñi JC (1958) Geología Uruguaya. Bol Inst Geol Uruguay 37:1–73

d´Orbigny A (1842a) Voyage dans l´Amérique Méridionale (le Brésil, la République Orientale de l´Uruguay, la République Argentine, la Patagonie, la République du Chili, la République de Bolivie, la République du Pérou), executé pendant années 1826–1833. 3(3) Géologie, P. Bertrand, Paris

d´Orbigny A (1842b) Voyage dans l´Amérique Méridionale (le Brésil, la République Orientale de l´Uruguay, la République Argentine, la Patagonie, la République du Chili, la République de Bolivie, la République du Pérou), executé pendant années 1826–1833. 3(4) Paléontologie, P. Bertrand, Paris

Darwin C (1839) Journal of researches into the geology and natural history of the various countries visited by H.M.S. Beagle. Colburn, London

Darwin C (1846) Geological observations on South America, being the third part of the geology of the voyage of the Beagle, under the command of capt. In: Fitzroy RN during the years 1832–1836. Smith, Elder and Co, London

Frenguelli JJ (1930) Apuntes de Geología Uruguaya. Inst Geol Perforaciones 11:1–47

Goso H (1965) El Cenozoico en el Uruguay. Inst Geol Uruguay 1–36

Goso H (1972) Cuaternario. Programa de Estudio y Levantamiento de Suelos del Ministerio de Ganadería, Agricultura y Pesca (MGAP), Montevideo, pp 1–12 (unpublished)

Goso H, Bossi J (1966) Cenozoico. In: Bossi J (ed) Geología del Uruguay. Div Publ Universidad de la República, Montevideo, pp 259–301

Jones G (1956) Mapa Geológico de la Región Oriental del Departamento de Canelones. Inst Geol Uruguay 34:1–193

Kraglievich L (1928) Apuntes para la Geología y Paleontología de la República Oriental del Uruguay. Rev Soc Amigos Arqueol 2:5–61

Kraglievich L (1932) Nuevos apuntes para la Geología y la Paleontología uruguayas. An Mus Hist Nat Montevideo 2:257–321

Lambert R (1941) Estado actual de nuestros conocimientos sobre la Geología de la República Oriental del Uruguay. Bol Inst Geol Uruguay 29:1–89

Larrañaga DA (1894) Memoria geológica sobre la formación del Río de la Plata, deducida de sus conchas fósiles, escrita por los años 1819. An Mus Nac Montevideo 1:3–12

Martínez S, Ubilla M (2004) El Cuaternario en Uruguay. In: Veroslavsky G, Ubilla M and Martínez S (eds) Cuencas sedimentarias de Uruguay. Geología, Paleontología y Recursos Naturales. Cenozoico, DIRAC-FC, Montevideo, pp 195–227

Martínez S, Ubilla M, Verde M, Perea D, Rojas A, Guerequiz R, Piñeiro G (2001) Paleoecology and geochronology of Uruguayan coastal marine Pleistocene deposits. Quat Res 55:246–254

Méndez-Alzola R (1950) Estudio sobre la obra científica de Larrañaga. Su iconografía paleomastozoológica. An Univ 60:49–129

Mones A (1981) Bibliografía sobre Geología y Paleontología del Uruguay. Bol Inst Geol Uruguay 38:95–162

Owen R (1837) A description of the cranium of the *Toxodon platensis*, a gigantic extinct mammiferous species, referrible by its dentition to the Rodentia, but with affinities to the Pachydermata and the Herbivorous Cetacea. Proc Geol Soc London 2:541–542

Paula Couto C (1948) Sobre os vertebrados fósseis da coleçaõ Sellow, do Uruguai. Dpto Nac Prod Min Div Geol Mineral RJ 125:1–14

Preciozzi F, Spoturno J, Heinzen W, Rossi P (1985) Carta Geológica del Uruguay a escala 1:500.000. Dirección Nacional de Minería y Geología, Montevideo

Serra N (1943) Memoria explicativa del mapa geológico del departamento de Colonia. Bol Inst Geol Uruguay 30:1–50

Serra N (1944) Memoria explicativa del mapa geológico del departamento de Treinta y Tres. Bol Inst Geol Uruguay 31:1–43

Serra N (1945) Memoria explicativa del mapa geológico del departamento de Soriano. Bol Inst Geol Uruguay 32:1–42

Teisseire A (1930) Contribución al estudio de la Geología y de la Paleontología de la República Oriental del Uruguay. Región de Colonia, Palacio del Libro, Montevideo

Veroslavsky G, Ubilla M, Martínez S (2004) Karl Walther. Primer Catedrático de Geología de Uruguay. In: Veroslavsky, Ubilla and Martínez (eds) Cenozoico. Cuencas Sedimentarias de Uruguay, DIRAC-FC, Montevideo, pp 357–383

Walther K (1915) Sobre la edad y carácter geológico de la Formación Pampeana. Sec Min Geol Agrol 1–35

Walther K (1919) Líneas fundamentales de la estructura geológica de la República Oriental del Uruguay. Rev Inst Nac Agron 3:1–186

Chapter 2
Marine and Transitional Deposits

Abstract Marine and transitional Quaternary deposits in Uruguay provide useful information about past sea levels and climatic and environmental changes over the last thousands of years. They yield abundant molluskan faunas that are crucial for these interpretations, and other invertebrates, vertebrates, ichnofaunas, microfaunas and microfloras. These deposits have been classified in a disorderly way over the years with a large number of stratigraphical units, but they can be reduced to two: the Pleistocene Chuy Formation, and the Holocene Villa Soriano Formation. They represent two main invasions of the sea, which displaced the Rio de la Plata estuary westward and northwestward. This phenomenon, and, logically, the rising sea level, was more intense in the Pleistocene. The fossil content not only gives proof of the relative temperature and salinity changes, but also of the level of the sea, and the age (since shells are very good material for radiocarbon dating). In the Late Pleistocene (MIS 5e) the sea level was about 6–8 m above the present level, and in the Holocene reached a maximum of 3 m above the present level, smoothly declining later. During both transgressions, the temperature was higher than at present, and faunal displacements have been verified; once again, these phenomena were clearly more intense in the Late Pleistocene. A considerable number of radiocarbon ages have been recorded, as well as some optically stimulated luminscence (OSL) ones.

Keywords Chuy Formation · Villa Soriano Formation · Sea level · Geology · Paleontology · Paleoecology · Radiocarbon · OSL/TL · Mollusks

2.1 Geological Features

Along the Uruguayan coast there are numerous outcrops of marine and transitional deposits that have been the subject of various approaches; unfortunately, most of them are considered "gray literature" with extremely local points of view. Due to space limitations, they will not be examined here.

A large number of lithostratigraphic or lithostratigraphic-like units have been proposed (Fig. 1.2), but only two are taken into account today: the Villa Soriano

© United Nations University 2016
M. Ubilla and S. Martínez, *Geology and Paleontology of the Quaternary of Uruguay*, SpringerBriefs in Earth System Sciences,
DOI 10.1007/978-3-319-29303-5_2

Formation (Goso 1972) (=Vizcaíno Gray Clays Caorsi and Goñi 1958), and the
Chuy Formation (Delaney 1963a, b) (*partim* Barra del Chuy Formation Bossi et al.
1998). Historical names can be consulted in Martínez et al. (2001, 2006) and
especially in the valuable Goñi and Hoffstetter publication (1964).

2.1.1 The Chuy Formation

Delaney (Delaney 1963a, b) defined the Chuy Formation in a Ph.D. thesis, but, in
fact, its definition was published for the first time by Goñi and Hoffstetter (1964),
who described the following lithology: quartz sands, reddish-yellow, and
semi-consolidated, with very rounded grains, polished, and sometimes mat bur-
nished, with an average diameter of 0.10–0.18 mm, and very well sorted.
Occasionally there are ferruginous nodules or incipient crusts. The sand is heavy in
minerals: zircon, tourmaline, biotite, rutile, garnet, andalusite, hornblende, epidote,
and actinolite; a ferruginous film usually covers the grains ("Arenas cuarzosas,
amarillo rojizas, semi-consolidadas, formada por granos bien redondeados, pulidos
y a veces de brillo mate, de diámetro medio de 0.10 a 0.18 mm, muy bien clasi-
ficados. Ocasionalmente se presentan nódulos ferruginosos o costras incipientes. La
arena posee minerales pesados: zircón, turmalina, biotita, rutilo, granate, andaluzita,
hornblenda, epidoto y actinolita; tiene los granos usualmente recubiertos por
película ferruginosa.").

Later usage widened this definition in order to include new deposits presumably
correlated (Bossi and Navarro 1991). For example, Preciozzi et al. (1985) added the
presence of gravel and clay lithologies to the Chuy Formation definition.

Bossi et al. (1998) proposed a name that was almost a homonym "Barra del
Chuy Formation," for the "sediments mapped as Chuy Formation by Elizalde
(1976), that correspond at least partially with the Paso del Puerto (Bossi et al. 1975)
and Chuy (Antón and Goso 1974) Formations)" ("los sedimentos cartografiados
como Formación Chuy por Elizalde (1976) y que se corresponden por lo menos
parcialmente con las denominaciones de Paso del Puerto (Bossi et al. 1975) y Chuy
(Antón and Goso 1974)"), with lithologies comprising from pelites to coarse
sandstones, determining a broadly coarsening up sequence.

It is clear that—excluding the original definition—the authors have privileged
the temporal (putative Pleistocene) and genetic (marine and marginal) components
of the deposits. Nevertheless, it is true that there are large lithological variations
when one moves laterally along the deposits, and these variations must be incor-
porated into a new, formal definition of the Chuy Formation. On the other hand, as
shown below, this wide lithological definition does not differ substantially from the
(also wide) characterization of the Holocene Villa Soriano Formation.

Indeed, the main (and probably only) difference between the Chuy and Villa
Soriano formations is their upper contact. The Chuy Formation would be overlayed

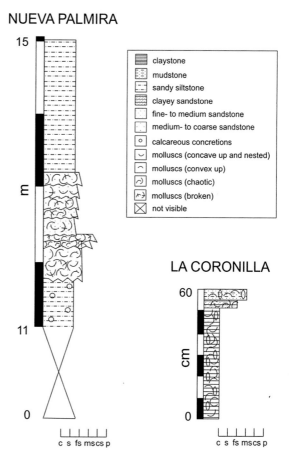

Fig. 2.1 Stratigraphic sections of the Chuy Formation (adapted from Martínez et al. 2001)

by Pleistocene continental units, and the Villa Soriano Formation would be over-layed only by recent deposits, or none at all. However, even this criterion is doubtful; as can be seen in Fig. 2.1, deposits undoubtedly belonging to the Pleistocene (La Coronilla) are at the top of the section. Other stratigraphic columns, including the Chuy Formation, can be seen in Goso Aguilar (2006) and Goso Aguilar et al. (2011).

Goso (1972), on the basis of the description of a few wells, postulated the existence of three marine episodes intercalated with continental deposits, named Chuy I, Chuy II, and Chuy III, respectively, linking them with the classic Alpine glaciations. This proposal has never been seriously evaluated, so it is now impossible to know whether it is reliable; what is clear, however, is that the use of lithostratigraphical names for chronostratigraphical units is problematic and brings only confusion to the literature.

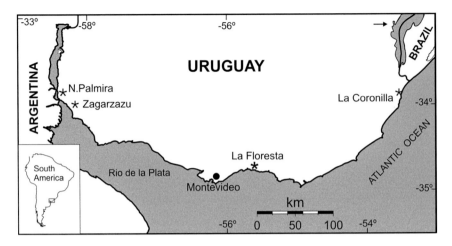

Fig. 2.2 Fossiliferous localities of the Chuy Formation

The Chuy Formation, in its broader sense, crops up all along the Uruguayan coastline; it was mapped by Preciozzi et al. (1985), and in more detail by Spoturno et al. (2004a, b, c) and Spoturno and Oyhantçabal (2012). In the chart by Bossi and Ferrando (2001), it is mentioned as the "Barra del Chuy" Formation. Figure 2.2 shows the location of the fossiliferous outcrops. As already stated by these authors, it is underlain by Proterozoic rocks, the continental Fray Bentos (Oligocene) and Libertad (Pleistocene) Formations, the marine Camacho Formation (Miocene), and overlain by the continental Pleistocene Dolores Formation and/or recent deposits. The thickness of the Chuy Formation is up to 30 m.

According to its stratigraphic relationships and numerical ages (see Sect. 2.2.1), the Chuy Formation (sensu lato) belongs to the Pleistocene. Its fossiliferous content (composed mostly of mollusks) also points in the same direction (see Sect. 2.3.1).

2.1.2 The Villa Soriano Formation

The fossiliferous deposits included in the Villa Soriano Formation (Goso et al., according to Antón and Goso 1974), crop out along the Uruguayan littoral, forming a stretch stripe parallel to the present coastline (broader at the mouths of rivers and coastal lagoons) (Fig. 2.3).

These deposits were previously called the "Vizcaíno Formation" (Goso and Bossi 1966), which was a nomenclatural upgrade of the "Arcillas grises de Vizcaíno" defined by Caorsi and Goñi (1958); this last name was at the same time a local name for the sedimentites previously known by the names of Argentinean origin "*querandinense*" and/or "*platense*."

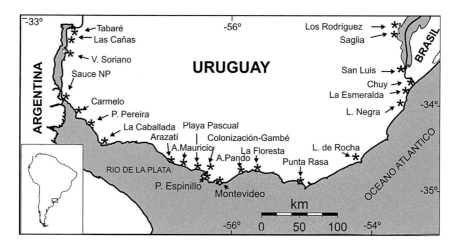

Fig. 2.3 Fossiliferous localities of the Villa Soriano Formation

Since its first mention, the definition of the Villa Soriano Formation was confusing. Caorsi and Goñi (1958) call it "gray clays" ("*arcillas grises*"), but stressed its fossiliferous (mollusks) content in the characterization of the unit, as did Goso and Bossi (1966) when they changed its name to the Vizcaíno Formation, highlighting the shell beds. Later, Preciozzi et al. (1985) stated that the Villa Soriano Formation comprises claystones, sandstones and conglomerates. To Bossi and Ferrando (2001), it is composed of very fine to coarse sands, some pelitic levels, and lenses of volcanic ash. Despite the great number of lithologies mentioned by the authors, and the different definitions of the unit, it is relatively easy to identify the Villa Soriano Formation in the field, mainly by the presence of marine and/or estuarine mollusks, which is a debatable lithological criterion, but, in practice, works reasonably well.

The Villa Soriano Formation overlies various geological units (Proterozoic rocks, Oligocene and Quaternary continental units, Fray Bentos and Libertad Formations) or Miocene marine sedimentites (Camacho Formation). On top, it has only recent sediments or soils, a fact that suggests a Holocene age, which is confirmed by the ^{14}C ages obtained (see Sect. 2.2.2).

There are two main groups of facies: a sandy one, represented by shoreface and foreshore deposits, and the other mostly pelitic, generated in environments of restricted water circulation, such as barrier-island-lagoon systems or very closed bays. The faunistic composition is concordant with these types, and in the pelitic deposits it is common to find bivalves in life position (see both facies groups in Figs. 2.4 and 2.5). Both facies are relatively thin; the thickness of the formation is, according to some authors, up to 20 m in subsurface (e.g., Spoturno et al. 2004a, b, c;

Fig. 2.4 Some outcrops of the Villa Soriano Formation (**a–b**) and shelly sandy facies (**e–f**); pelitic facies with *T. plebeius* in life position (**c–d**)

Spoturno and Oyhantçabal 2012), but the outcrops are usually not thicker than 5 m. Due to its thinness and extension, this unit is mapped in charts only at 1:100,000 or lower (see, for example, Spoturno et al. 2004a, b, c; Spoturno and Oyhantçabal 2012).

The regional equivalents of the Villa Soriano are the Patos Formation Group in Rio Grande do Sul (Brazil) (Forti Esteves 1974; Martínez 1990), and the "*querandinense*," and "*platense*" deposits of Argentina (presently with several local names; see Martínez et al. 2006, 2013).

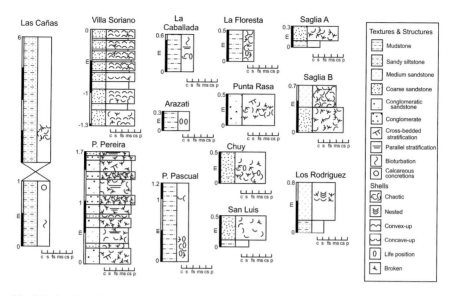

Fig. 2.5 Stratigraphic sections of the Villa Soriano Formation (adapted from Martínez et al. 2006)

2.2   Numerical Ages

2.2.1   The Chuy Formation

The Pleistocene deposits have been dated with the conventional ^{14}C, putting them between ca. 30,000 and ca. 35,000 years BP (Martínez et al. 2001; Rojas 2007) (Table 2.1). They were interpreted by Martínez et al. (2001) as minimum ages, and the deposits were considered to belong to the last interglacial period identified with

Table 2.1 Numerical ages from the Pleistocene of Uruguay

Lab. ID	Taxon sample	M	SL	Me	^{14}C age BP
LP-738	*Mactra isabelleana*	Shell	NP	^{14}C	31,000 ± 1,200
LP-730	*Anomalocardia brasiliana*	Shell	NP	^{14}C	34,600 ± 2,000
LP-884	*Ostrea equestris*	Shell	LC	^{14}C	29,500 ± 600
LP-824	*Mactra isabelleana*	Shell	LC	^{14}C	35,500 ± 1,900
LP-1466	*Tagelus plebeius*	Shell	Z	^{14}C	35,500 ± 1,900
UIC2632	–	Sand	NP	OSL	80,680 ± 5,500
UIC2633	–	Sand	Z	OSL	88,355 ± 7,070

^{14}C is conventional in all cases (Martínez et al. 2001; Rojas 2007, 2010)
M material, *SL* sample location, *Me* Method, *NP* Nueva Palmira, *LC* La Coronilla, *Z* Zagarzazú, *OSL* Optically Stimulated Luminescence

the Marine Isotope Stage 5e (MIS5e), attending to their malacological content. Optically Stimulated Luminescence (OSL) dating, published by Rojas in 2010, gave them ages of between ca. 80,000 and 88,000 years (Table 2.1), but they have not yet been sufficiently discussed.

2.2.2 The Villa Soriano Formation

As previously mentioned, the Villa Soriano Formation, according to the ^{14}C ages, is dated as Holocene. There is now a large number of ^{14}C ages mentioned that were obtained from various mollusk species and paleoenvironments (Table 2.2 and references herein), which vary from ca. 7,000 to 1,500 ^{14}C years, and between these extremes are marine or estuarine deposits all along the period. According to Cortelezzi and Lerman (1971), Delaney (1967) gave some ^{14}C ages for Uruguay, which would be the first obtained for the country.

2.3 Paleontological Content

The Quaternary of Uruguay has abundant marine and estuarine fossils—mainly mollusks—but also other invertebrates, vertebrates, microfossils, and trace fossils. By far, the most abundant and studied are the mollusks. Clavijo et al. (2005) provide a list of the first mentions of the Quaternary of Uruguay, without distinguishing between Pleistocene or Holocene ones. Before Martínez et al. (2001), most paleontological studies were age constrained only by the stratigraphic position of the outcrops or wells, their self-fossiliferous content, and even by their appearance. As a consequence, older paleontological references must to be checked to know—in light of the numerical ages—whether the cited taxa belong to the Pleistocene, Holocene, or if they are merged, and it is not possible to differentiate between them; such is the case of the global list of Clavijo et al. (2005), or in wells never dated, such as those studied by Sprechmann (1978). The following considerations take into account only reliable data.

2.3.1 The Chuy Formation

The mollusks (around 40 species) can be consulted in detail in Martínez et al. (2001, 2013), in particular for bivalves and gastropods. Rojas and Urteaga (2011)

Table 2.2 ^{14}C conventional and AMS ages from the Villa Soriano Formation

Lab ID	TS	M	SL	^{14}C age BP	Cal BP
AA93722*	1	Shell	CPolonio1	462 ± 37.1	0–148
URU0056	2	Shell	LCastillos1	1,480 ± 52.8	913–1,152
URU417	2	Shell	LCCordonGM	1,680 ± 52.8	1,106–1,331
URU0408	2	Shell	LCCordonGM	1,700 ± 52.8	1,130–1,350
URU0412	2	Shell	LCCordonGM	1,770 ± 52.8	1,200–1,430
LP-1365	2	Shell	LCastillos5	1,810 ± 141	1,059–1,679
URU0401	2	Shell	LCC3Silva3	2,040 ± 62.4	1,432–1,768
URU0398	2	Shell	LCC2Cnegro	2,070 ± 72	1,455–1,819
AA93853*	3	Shell	Esmeraldainf	2,100 ± 39.8	1,552–1,790
URU0400	2	Shell	LCC3Silva	2,190 ± 62.4	1,608–1,927
URU0060	2	Shell	LCastillos2	2,220 ± 81.8	1,596–2,009
AA90343*	4	Shell	LGarzon1	2,290 ± 43.5	1,775–2,013
URU310	2	Shell	Colonizacion1	2,310 ± 62.4	1,764–2,092
AA86683*	2	Shell	Gambe1	2,365 ± 36.2	1,878–2,093
URU0236	U	Shell	Esmeralda8	2,370 ± 52.8	1,856–2,127
URU0057	2	Shell	LCC2Silva	2,480 ± 62.4	1,967–2,298
LP-817	5	Shell	PuntaFria	2,490 ± 72	1,957–2,314
URU552	2	Shell	Colonizacion2	2,600 ± 52.8	2,115–2,395
URU551	2	Shell	Colonizacion3	2,620 ± 52.8	2,136–2,435
URU309	2	Shell	Gambe2	2,710 ± 62.4	2,281–2,656
AA86685*	2	Shell	Colonizacion4	2,744 ± 52.8	2,323–2,652
AA93720*	4	Shell	PuntaRasa	2,875 ± 86.7	2,350–2,785
URU0059	2	Shell	LCastillos3	2,930 ± 72	2,462–2,852
URU0416	2	Shell	LCC1Silva2	2,990 ± 52.8	2,661–2,890
AA90345*	6	Shell	LGarzon2	3,077 ± 39.8	2,743–2,943
URU0235	U	Shell	Esmeralda9	3,240 ± 62.4	2,867–3,233
LP-753	2	Shell	PPereira2	3,300 ± 62.4	2,952–3,312
AA90346*	7	Shell	PdelEsteM1	3,301 ± 39.8	2,991–3,264
AA83199*	8	Shell	PEspinillo1	3,501 ± 43.5	3,255–3,483
LP-907	2	Shell	SagliaA	3,530 ± 52.8	3,286–3,553
LP-775	2	Shell	PPereira1	3,590 ± 62.4	3,336–3,631
LP-1375	2	Shell	Carmelo	3,710 ± 62.4	3,457–3,803
URU009	9	Shell	PEspinillo	3,790 ± 141	3,374–4,090
AA88096*	5	Shell	Pcarretas	3,815 ± 61	3,678–3,844
URU0077	2	Shell	Conchillas	3830 ± 52.8	3,622–3,920
LP-887	2	Shell	LosRodriguez	4,000 ± 72	3,810–4,221
AA93852*	3	Shell	Valizas1	4,092 ± 42.5	3,976–4,265
AA88094*	4	Shell	PdelEste2	4,312 ± 56.6	4,245–4,588
URU0131	14	Shell	Valizas2	4,360 ± 62.4	4,300–4,701
URU0058	14	Shell	Valizas3	4,370 ± 62.4	4,336–4,721
URU0063	2	Shell	Castro4	4,410 ± 52.8	4,411–4,730
LP-893	2	Shell	SagliaB2	4,460 ± 72	4,436–4,810

(continued)

Table 2.2 (continued)

Lab ID	TS	M	SL	^{14}C age BP	Cal BP
AA93721*	10	Shell	PLecoq	4,481 ± 41.6	4,529–4,796
LP-913	2	Shell	LasCañas	4,550 ± 62.4	4,538–4,874
URU 0398	11	Shell	LCCnegro	4,620 ± 62.4	4,639–5,008
URU 0396	11	Shell	LCCServeto	4,640 ± 72	4,624–5,055
URU0159	14	Shell	AoMauricio	4,750 ± 72	4,828–5,230
LP-904	3	Shell	LaFloresta	4,790 ± 81.8	4,846–5,271
LP-901	2	Shell	SagliaB1	4,800 ± 72	4,864–5,268
URU0006	2	Shell	Saglia	4,810 ± 141	4,725–4,748
URU008	2	Shell	LMSLuis	4,830 ± 151	4,727–4,747
URU0064	2	Shell	Castro2	4,870 ± 62.4	4,956–5,314
LP-1372	2	Shell	Tabare	4,930 ± 72	5,022–5,445
LP-821	3	Shell	Chuy	5,070 ± 72	5,278–5,573
URUAC1206	2	Shell	LMSantos	5,090 ±111.3	5,131–5,160
URU0308	3	Shell	Esmeralda11	5,110 ± 62.4	5,310–5,578
LP-829	2	Shell	SanLuis	5,150 ± 81.8	5,303–5,652
URU0065	2	Shell	Castro1	5,210 ± 62.4	5,433–5,707
AA81058*	14	Shell	AoCeibos	5,280 ± 72	5,476–5,814
URU0055	2	Shell	LCastillos4	5,287 ± 45.3	5,556–5,747
URU0014	2	Shell	CMorgan1	5,480 ± 72	5,652–6,005
URU0015	2	Shell	CMorgan2	5,520 ± 72	5,712–6,095
URU177	2	Shell	NBerlin	5,520 ± 72	5,712–6,095
AA88093*	14	Shell	AoSarandi1	5,562 ± 45.3	5,845–6,097
AA92198*	12	Shell	IsladeTala	5,691 ± 45.3	5,953–6,207
AA88350*	13	Shell	SauceNP1	5,714 ± 64.3	5,954–6,261
AA88351*	13	Shell	SauceNP2	5,728 ± 55.7	5,985–6,262
LP-740	2	Shell	VSoriano2	5,840 ± 72	6,064–6,072
URU0072	2	Shell	VSoriano4	5,850 ± 62.4	6,122–6,396
LP-815	3	Shell	P.Pascual	5,870 ± 91.6	6,054–6,483
URU0069	2	Shell	VSoriano3	5,910 ± 62.4	6,190–6,449
LP-806	3	Shell	LaCaballad	6,020 ± 86.7	6,265–6,633
LP-747	10	Shell	Arazati	6,260 ± 91.6	6,477–6,930
AA88091*	14	Shell	AoSanGregorio	6,271 ± 53.8	6,579–6,866
AA81057*	3	Shell	StaAna	6,589 ± 48.1	6,970–7,228
LP-1464	13	Shell	AoPando	6,630 ± 62.4	6,978–7,280
UG11502c*	15	Shell	ASolis	7,160 ± 30	7,860–8,011

(Beovide 2011; Bracco and Ures 1998; Bracco et al. 1994, 2000, 2003 2011; Martínez et al. 2006; Martínez and Rojas 2013; Mata et al. 2012; Rojas 2007; Rojas and Martínez 2004; Rojas and Urteaga 2011; Mourelle et al. 2015)

*AMS ones

TS taxon sample, *U* undetermined, *M* material, *SL* sample location, *1 Urosalpinx haneti, 2 Erodona mactroides, 3 Mactra isabelleana, 4 Glycymeris longior, 5 Tegula patagonica, 6 Crepidula protea, 7 Lottia subrugosa, 8 Notocochlis isabelleana, 9 Ostrea* sp., *10 Ostreola equestris, 11 Amiantis purpurata, 12 Chione cancellata, 13 Anomalocardia brasiliana, 14 Tagelus plebeius, 15 Heleobia australis*

describe polyplacophorans, and other less represented taxa can be found in Rojas and Martínez (in press). (A full list of mollusks is given in Appendix 1.) The most common species are the bivalves *Mactra isabelleana*, *Ostrea equestris*, *Anomalocardia brasiliana*, *Caryocorbula caribaea*, *Pitar rostratus*, and the gastropods *Bostrycapulus odites* and *Stramonita haemastoma* (Martínez et al. 2001) (Fig. 2.6). All taxa known for the Chuy Formation are extant.

Few trace-fossils are referred to in this unit (Fig. 2.7). Goso Aguilar (2006) mentioned *Ophiomorpha*, and Rojas et al. (2014) described *Oichnus*, which is particularly important because it represent the only predatory trace found on plates of chitons.

2.3.2 The Villa Soriano Formation

Holocene mollusks are listed in Figueiras (1961, 1962, 1967, with the last including mentions of other taxa), Martínez et al. (2006, 2013, bivalves and gastropods), and Rojas and Urteaga (2011, polyplacophorans). A list of them is in Appendix 2. The most frequently found species depend on the facies and salinity involved. In estuary environments, the bivalves *Erodona mactroides* and *Mactra isabelleana* (sandy substrates), *Tagelus plebeius* (pelitic substrates), and *Ostrea equestris* are abundant, as are the gastropods *Heleobia australis* and *H. charruana*. In polyhalyne and euhahlyne environments, the bivalves *Mytilus edulis* (in the vicinity of rocky substrates), *Glycymeris longior*, *Anadara brasiliana*, *Noetia bisulcata*, *Mactra* spp., and the gastropods *Buccinanops* spp. and *Urosalpinx haneti* are frequently found (Fig. 2.8).

Other Holocene taxa, which are simply mentioned in most of the cases, are: benthonic foraminifers (Sprechmann 1978), diatoms (Frenguelli 1930; Perea 1981; Larrañaga 1990; Metzeltin and García-Rodríguez 2003), phytoliths (del Puerto et al. 2013 and references therein), silicoflagelates (Perea 1981), briozoans (Figueiras and Broggi 1967), decapod crustaceans (Figueiras and Broggi 1967; Martínez 1983; Verde et al. 1998), cirripeds (Figueiras and Broggi 1967), Cnidaria (Figueiras and Broggi 1967), serpulids (Figueiras and Broggi 1967), sponge spicles (Frenguelli 1930; Perea 1981; Larrañaga 1989), echinoderms (Figueiras and Broggi 1967; Martínez and Mooi 2005), chondrichtyes and osteichthyes (Perea and Ubilla 1981; Piñeiro and Verde 1999; Rinderknecht et al. 2005), and aves (Mones 1993).

Trace-fossils have also been described; they are mainly from hard substrates (Verde 2003; Lorenzo and Verde 2004) (Fig. 2.7). Appendix 3 contains a list of non-molluskan fossils from the Villa Soriano Formation. All of them, as well as mollusks, are extant.

Fig. 2.6 The most common molluskan species of the Chuy Formation: **a–b** *Anomalocardia brasiliana* (*internal and external views*), **c** *Caryocorbula caribaea* (*internal view*), **d–e** *Ostrea equestris* (*internal and external views*), **f–g** *Bostrycapulus odites* (*internal and external views*), **h–i** *Mactra isabelleana* (*internal and external views*), **j–m** *Stramonita haemastoma* (*ventral and dorsal views*), **k–l** *Pitar rostratus* (*internal and external views*). *Scales* 10 mm

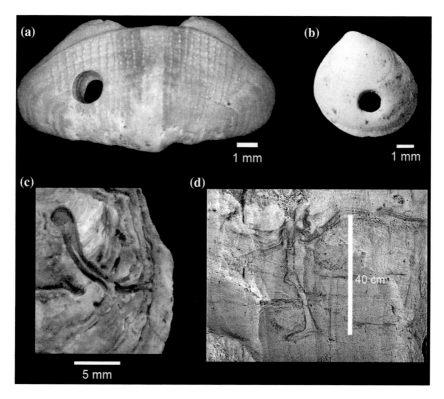

Fig. 2.7 Some ichnofossils from the Chuy and Villa Soriano Formations: **a** *Oichnus simplex* on *Chaetopleura angulata* (Chuy Formation) (*photo* Alejandra Rojas), **b** *Oichnus* on *Glycymeris longior* (Villa Soriano Formation), **c** *Caulostrepsis* sp. on *Ostrea equestris* (Villa Soriano Formation) (*photo* Mariano Verde), **d** *Thalassinoides* sp. (Chuy Formation) (*photo* César Goso)

Bedani et al. (2000) indicate in an abstract some microfossil (mainly palyno-logical) content of three pelitic deposits, two Holocene and one known as Pleistocene, but this content is not differentiated among the localities. The groups mentioned are chitinous foraminifers, dinoflagellates, briophyte and pteridophyte spores, and pollen of Amaranthaceae-Chenopodioideae. Mourelle et al. (2015) studied the diatom, pollen, and others from the marshes of the Solís Grande Creek, adding a great number of new references for the Holocene of Uruguay. Also mentioned in the literature are some terrestrial mollusks (*Megalobulimus*, *Austroborus*, Gastrocopta, *Radiodiscus*, *Zylchogira*, *Bulimulus*—see Figueiras and Broggi 1967), which are surely allochthonous and transported by fluvial or even pluvial waters to the coastal deposits.

Fig. 2.8 Frequent mollusk species of the Villa Soriano Formation: **a–b** *Erodona mactroides* (external and internal views), **c–d** *Urosalpinx haneti* (ventral and dorsal views), **e–f** *Mactra isabelleana* (external and internal views), **g–h** *Buccinanops cochlidium* (ventral and dorsal views), **i–j** *Mytilus edulis* (internal and external views), **k** *Heleobia* sp. (ventral view), **l** *Ostrea equestris* (internal view), **m–n** *Noetia bisulcata* (external and internal views), **o–p** *Tagelus plebeius* (internal and external views). *Scales* 5 mm (except k)

2.4 Environment and Climatic Scenarios

2.4.1 Salinity and Temperature

The paleoecology of the molluskan faunas and its relationship with the depositional environments were studied mostly in the last decades. As expected, a westward displacement of the salinity gradient was found, more prevalent in the Pleistocene than in the Holocene; this is in consonance with the magnitude of the respective transgressions, with a higher sea level in the Pleistocene (MIS5e). During this lapse, we found mollusks that indicate euhaline-polyhaline conditions in the western localities (e.g., Nueva Palmira); meanwhile in the Holocene in this zone, mixohaline conditions prevailed (Figs. 2.9, 2.10 and 2.11) (Martínez et al. 2001, 2006).

All the mollusk assemblages show that the water temperature was warmer than it is today—again, more intensely in the Late Pleistocene; this is evidenced by the presence of species that today are not found on the Uruguayan coast, but northwards, where the water is warmer. In other words, their southern distributional limits have shifted to the north. It is also confirmed by the proportion of species with warm- or cold-water affinities, with the former being more abundant in the Quaternary assemblages than they are today; on the contrary, there are very few, or even no cold-water species in the Quaternary assemblages (Figs. 2.12 and 2.13). The Pleistocene species with present southern distributional limit today northwards than in the Pleistocene are *Finella dubia*, *Chrysallida gemmulosa*, *Chione subrostrata*, *Anomalocardia brasiliana*, *Laevicardium* sp., *Limaria* sp., and in the Holocene *Anomalocardia brasiliana*, *Marshallora nigrocinta*, *Turbonilla abrupta*, *Cerithiopsis greeni* and *Seila adamsii* (Martínez et al. 2013). The evolution of

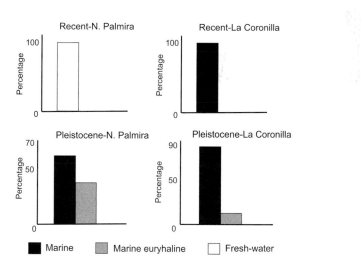

Fig. 2.9 Salinity in Pleistocene localities according to the mollusks (adapted from Martínez et al. 2001)

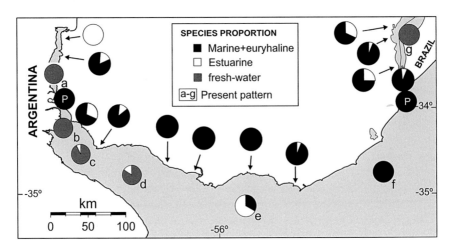

Fig. 2.10 Salinity in Holocene localities according to the mollusks (Martínez et al. 2006)

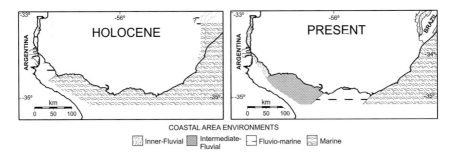

Fig. 2.11 Salinity environments on the Uruguayan coast, Holocene and today (Martínez et al. 2006)

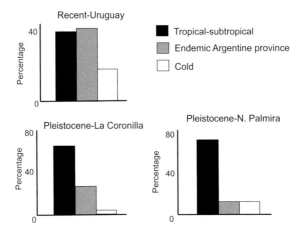

Fig. 2.12 Temperatures in Pleistocene localities according to the mollusks (adapted from Martínez et al. 2001)

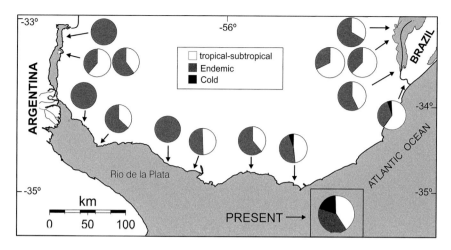

Fig. 2.13 Temperatures in Holocene localities according to the mollusks (Martínez et al. 2006)

special environments has also been studied, such as lakes (e.g., Garcia-Rodriguez et al. 2010; del Puerto et al. 2013, and references therein), and marshes (Mourelle et al. 2015).

2.4.2 Sea Level

The sea level has not been precisely established for the Pleistocene deposits, although it is expected that in a passive continental margin it follows the global trend. In this sense, at Nueva Palmira, a foreshore (wave-dominated) deposit, the sea level may have reached 6–8 m above the present one, but the Zagarzazú (Colonia Department) and La Coronilla (Rocha Department) seem to represent lower levels. It is also possible that there have been some neotectonic processes (not discarded by Martínez et al. 2001), or that the sites are not strictly contemporaneous.

Bracco-Boksar et al. (2011), combining evidence with varying reliability, and not adjusting for the height of the deposition, constructed by hand a curve based on the deposits around the current Rocha Lagoon, and concluded that the maximum Holocene sea level was reached approximately 6,000 years ago, and that since then there has been a fluctuating declining trend. According to Martínez and Rojas (2013), who statistically used data from old berm deposits corrected for the height of the deposition, the maximum sea level (around 3 m above the present sea level) was reached also around 6,000 years ago, but that after that there was a smooth lowering of the sea level (Fig. 2.14) (see also Bracco et al. 2014; Martínez and Rojas 2014; Mourelle et al. 2015).

Fig. 2.14 Sea level in the
Holocene (Martínez and
Rojas 2013)

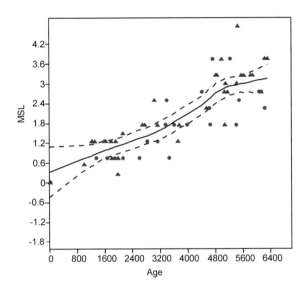

References

Antón D, Goso H (1974) Estado actual sobre el conocimiento del Cuaternario en el Uruguay. Dirección de Suelos y Fertilizantes, Montevideo

Beovide L (2011) Arqueozoología de los Depósitos Conchilíferos de la Cuenca Inferior del Río Santa Lucía. Dissertation, Universidad de la República, Uruguay

Bossi J, Ferrando L (2001) Carta geológica del Uruguay. Escala 1:500.000. V. 2.0. Montevideo, Geoeditores

Bossi J, Navarro R (1991) Geología del Uruguay. Departamento de Publicaciones, Universidad de la República, Montevideo

Bossi JC, Ferrando LA, Fernández N, Elizalde G, Morales H, Ledesma JJ, Carballo E, Medina E, Ford I, Montaña JR (1975) Carta Geológica del Uruguay, Escala 1:1.000.000. Dirección de Suelos y Fertilizantes, MAP, Montevideo

Bossi J, Ferrando LA, Montaña J, Campal N, Morales H, Gancio F, Schipilov A, Piñeyro D, Sprechmann P (1998) Carta Geológica del Uruguay, Escala 1/500.000. Geoeditores SRL, Montevideo

Bracco R (2003) Aproximación al registro arqueológico del sitio La Esmeralda ("Conchero"), desde su dimensión temporal. Costa atlántica del Uruguay. An Arqueol Etnol 54–55:13–26

Bracco R, Ures C (1998) Las variaciones del nivel del mar y el desarrollo de las culturas prehistóricas del Uruguay. Actas II Congreso Uruguayo de Geología:16–20

Bracco R, Navarro R, Ures C (1994) Primer intento cronológico de una ingresión holocénica en Carrasco (Montevideo, Uruguay). Paper presented at the 1º Jornada de Paleontología del Uruguay, Montevideo, 2–6

Bracco R, Panario D, Ures C (1999) Dataciones de 14C y efecto de reservorio para el litoral del Uruguay. Paper presented at the I Jornadas del Cenozoico del Uruguay, Montevideo, pp 4–5

Bracco R, Montaña J, Bossi J, Panarello H, Ures C (2000) Evolución del humedal y ocupaciones humanas en el sector sur de la cuenca de la laguna Merín. In: Durán A, Bracco R (eds) Arqueología de las Tierras Bajas. Ministerio de Educación y Cultura, Montevideo, pp 99–116

Bracco R, Inda H, del Puerto L, Capdepont I, Panario D, Castiñeira C, García-Rodríguez F (2014) A reply to "Relative sea level during the Holocene in Uruguay". Palaeogeogr Palaeoclimatol Palaeoecol 401:166–170

Bracco-Boksar R, García-Rodríguez F, Inda H, del Puerto L, Castiñeira C, Panario D (2011) Niveles relativos del mar durante el Pleistoceno final-Holoceno en la costa de Uruguay. In: García-Rodríguez F (ed) El Holoceno en la zona costera de Uruguay. Dpto. Publicaciones UDELAR, Montevideo, pp 65–92

Caorsi J, Goñi J (1958) Geología Uruguaya. Bol Inst Geol Uruguay 37:1–73

Clavijo C, Scarabino F, Rojas A, Martínez S (2005) Lista sistemática de los moluscos marinos y estuarinos del Cuaternario del Uruguay. Com Soc Malac Uruguay 9:381–411

Cortelezzi CR, Lerman JC (1971) Estudio de las formaciones marinas de la costa Atlántica de la provincia de Buenos Aires. An LEMIT 1(133):164

de Bedani EF, Garcia MJ, Veroslavsky G, Martínez S (2000) Resultados palinológicos preliminares da Formação Villa Soriano, Quaternário do Uruguai. Rev Univ Guarulhos (Geociências) V(N Especial):273

del Puerto L, Bracco R, Inda H, Gutiérrez O, Panario D, García-Rodríguez F (2013) Assessing links between late Holocene climate change and paleolimnological development of Peña Lagoon using opal phytoliths, physical and geochemical proxies. Quat Res 287:89–100

Delaney PJV (1963) Quaternary Geologic History of the coastal plain of Rio Grande do Sul, Brazil. Lousiana State Univ Coastal studies series 7:1–63

Delaney PJ (1967) Geomorphology and Quaternary Coastal geology of Uruguay. Edited by the author: Fide Cortelezzi and Lerman (1971)

Elizalde G (1976) Conservación y mejora de playas. Informe Ministerio de Transporte y Obras Públicas - UNESCO URU-73-007. Ed. UNESCO, Montevideo, Uruguay. Fide Bossi et al. (1998)

Figueiras A (1961) Contribución al conocimiento de la malacofauna holocena del Uruguay. Com Soc Malac Uruguay 1:15–21

Figueiras A (1962) Sobre nuevos hallazgos de moluscos subfósiles de la Transgresión Querandina. Com Soc Malac Uruguay 1:53–68

Figueiras A (1967) Contribución al conocimiento de los moluscos del Holoceno uruguayo. Com Soc Malac Uruguay 2:61–76

Figueiras A, Broggi J (1967) Estado actual de nuestros conocimientos sobre los moluscos fósiles del Uruguay Parte I. Com Soc Malac Uruguay 2:147–187

Forti Esteves IR (1974) Bioestratigrafia e palaeoecologia (Mollusca) do Quaternario da Planicie costeira do Rio Grande do Sul (Brasil). Ann XXVIII Congreso Brasileiro Geología 133–149

Frenguelli J (1930) Apuntes de Geología uruguaya. Bol Inst Geol Perf 11:1–47

Garcia-Rodriguez F, Stutz S, Inda H, Del Puerto L, Bracco R, Panario D (2010) A multiproxy approach to infer Holocene paleobotanical changes linked to sea-level variation, paleosalinity levels and shallow lake alternative states in Negra Lagoon, SE Uruguay. Hydrobiologia 646:5–20

Goñi JC, Hoffstetter R (1964) Uruguay. Lexique Stratigraphique. International, 5 Amérique Latine, 9a. CNRS, Paris

Goso H (1972) Cuaternario. Programa de Estudio y Levantamiento de Suelos del Ministerio de Ganadería, Agricultura y Pesca (MGAP), Montevideo, pp 1–12 (unpublished)

Goso Aguilar C (2006) Aspectos sedimentológicos y estratigráficos de los depósitos cuaternarios de la costa platense del Departamento de Canelones (Uruguay). Lat Amer J Sed Basin An 13:77–89

Goso Aguilar C, Mesa V, Alvez MC (2011) Sinopsis geológico-ambiental de la costa platense y atlántica de Uruguay. In: López RA, Marcomini SC (eds) Problemática de los ambientes costeros. Sur de Brasil, Uruguay y Argentina. Croquis, Buenos Aires, pp 59–76

Goso H, Bossi J (1966) Cenozoico. In: Bossi J (ed) Geología del Uruguay. Div Publ Universidad de la República, Montevideo, pp 259–301

Larrañaga RJ (1989) Listado específico de las esponjas (Porifera) actuales y fósiles del Uruguay y nueva propuesta sistemática. Bol Soc Zool Uruguay 2(5):51

Larrañaga RJ (1990) Nuevos aportes micropaleontológicos (Bacillariophyta) en sedimentos platenses (Fm. Villa Soriano, Holoceno). Paper presented at the 1st Congreso Uruguayo de Geología 2, Montevideo

Lorenzo N, Verde M (2004) Estructuras de bioerosión en moluscos marinos de la Formación Villa
    Soriano (Pleistoceno tardío—Holoceno) de Uruguay. Rev Bras Paleont 7:319–328
Martínez S (1983) Invertebrados fósiles holocenos del Departamento de Rocha, Uruguay
    (Mollusca y Crustacea). Res Com Jorn Cienc Nat 3:46–47
Martínez S (1990) Taphonomy and Palaeoecology of Holocene mollusks from the western margin
    of the Merin Lagoon (Villa Soriano Fm., Uruguay). Quat South Amer Antarct Pen 7:121–135
Martínez S, Mooi R (2005) Extinct and extant sand dollars (Clypeasteroida: Echinoidea) from
    Uruguay. Rev Biol Trop/Int J Trop Biol 53 Suppl 3:1–7
Martínez S, Rojas A (2013) Relative sea level during the Holocene in Uruguay. Palaeogeogr
    Palaeoclimatol Palaeoecol 374:123–131
Martínez S, Rojas A (2014) Comments to "A reply to Relative sea level during the Holocene in
    Uruguay" by Bracco et al. (2013). Palaeogeogr Palaeoclimatol Palaeoecol 401:171–172
Martínez S, Ubilla M, Verde M, Perea D, Rojas A, Guérèquiz R, Piñeiro G (2001) Paleoecology
    and geochronology of Uruguayan coastal marine Pleistocene deposits. Quat Res 55:246–254
Martínez S, Rojas A, Ubilla M, Verde M, Perea D, Piñeiro G (2006) Molluscan assemblages from
    the marine Holocene of Uruguay: composition, geochronology and paleoenvironmental
    signals. Ameghiniana 43:385–398
Martínez S, del Río CJ, Rojas A (2013) Biogeography of the quaternary molluscs of the
    Southwestern Atlantic ocean. Springer briefs in earth system sciences. Springer, Heidelberg
Mata V, Arruabarrena Ottati A, Martínez S (2012) Un camino de gestión patrimonial y producción
    de conocimiento. El fuerte San José como caso de investigación patrimonial. Rev Arqueol Hist
    Arg Latinoamer 6:63–94
Metzeltin D, García-Rodríguez F (2003) Las Diatomeas uruguayas. DIRAC, Montevideo
Mones A (1993) Notas paleontológicas uruguayas. V. Primer hallazgo de *Rhea* sp. fósil en el
    Uruguay (Aves: Struthioniformes: Rheidae). Com Paleont Mus Nat 2:65–68
Mourelle D, Prieto AR, Pérez L, García-Rodríguez F, Borel CM (2015) Mid and late Holocene
    multiproxy analysis of environmental changes linked to sea-level fluctuation and climate
    variability of the Río de la Plata estuary. Palaeogeogr Palaeoclimatol Palaeoecol 421:75–88
Perea D (1981) Algunos microfósiles silíceos (Bacillariophyceae, Silicoflagellata, Porifera)
    encontrados en el Platense de las cercanías de Nueva Palmira. Dpto. de Colonia. Dissertation,
    Udelar
Perea D, Ubilla M (1981) Estudio preliminar de la ictiofauna fósil marina de las costas del Dpto.
    de Rocha, Uruguay. Res Com Jorn Cienc Nat Uruguay 2:25–26
Piñeiro G, Verde M (1999) Osteología y paleoecología de Sciaenidos (Teleostei, Perciformes) de
    la Formación Villa Soriano (Holoceno) de Uruguay. Rev Univ Guarulhos Geociências
    4(6):19–27
Preciozzi F, Spoturno J, Heinzen W, Rossi P (1985) Carta Geológica del Uruguay a escala
    1:500.000. Dirección Nacional de Minería y Geología, Montevideo
Rinderknecht A, Piñeiro G, Verde M, Rojas A (2005) New fish remains from marine deposits of
    Uruguay. Paper presented at the VIII Jornadas de Zoología del Uruguay, Montevideo
Rojas A (2007) Moluscos de aguas cálidas del Cuaternario Marino del Uruguay. M.Sc.
    Dissertation, PEDECIBA, Universidad de la República
Rojas A (2010) Primeras edades obtenidas por datación por luminiscencia (OSL) en depósitos
    marinos fosilíferos del Pleistoceno tardío de Uruguay. Paper presented at the VII Congreso
    Latinoamericano de Paleontología, X Congreso Argentino de Paleontología y Bioestratigrafía,
    Universidad de la Plata, La Plata
Rojas A, Martinez S (in press) MIS 3 vs. MIS 5 fossiliferous marine deposits from Uruguay. In:
    Gasparini GM, Rabassa J, Deschamps MC, Tonni, EP (eds) Marine Isotope Stage 3 in
    Southern South America 60 KA B.P.-30 KA B.P. Springer Earth System Sciences
Rojas A, Martínez S (2004) Nuevas dataciones radiocarbónicas para el Cuaternario marino de
    Uruguay. Paper presented at the IV Congreso Uruguayo de Geología, Montevideo
Rojas A, Urteaga D (2011) Late Pleistocene and Holocene chitons (Mollusca, Polyplacophora)
    from Uruguay: palaeobiogeography and palaeoenvironmental reconstruction in mid latitudes of
    the Southwestern Atlantic. Geobios 44:377–386

Rojas A, Verde M, Urteaga D, Scarabino F, Martínez S (2014) The first predatory drillhole on a fossil chiton plate: an occasional item prey or an erroneous attack? Palaios 29:414–419

Spoturno J, Oyhantçabal P (2012) Mapa Geológico y de Recursos Minerales del Departamento de Maldonado. Montevideo, DINAMIGE-Facultad de Ciencias

Spoturno J, Oyhantçabal P, Goso C, Aubet N, Cazaux S, Huelmo S, Morales E, Loureiro J (2004a) Mapa geológico del Departamento de Montevideo a escala 1:50.000. Memoria descriptiva + mapa, CD. CONICYT—Fac. Ciencias–DINAMIGE, Montevideo

Spoturno J, Oyhantçabal P, Goso C, Aubet N, Cazaux S, Huelmo S, Morales E, Loureiro J (2004b) Mapa geológico del Departamento de Canelones a escala 1:100.000. Memoria descriptiva + mapa, CD. CONICYT—Fac. Ciencias–DINAMIGE, Montevideo

Spoturno J, Oyhantçabal P, Goso C, Aubet N, Cazaux S, Huelmo S, Morales E, Loureiro J (2004c) Mapa geológico del Departamento de San José a escala 1:100.000. Memoria descriptiva + mapa, CD. CONICYT—Fac. Ciencias–DINAMIGE, Montevideo

Sprechmann P (1978) The paleoecology and paleogeography of the Uruguayan coastal area during the Neogene and Quaternary. Zitteliana 4:3–72

Verde M (2003) Marine invertebrate ichnofossils from the Quaternary of Uruguay. Paper presented at the VII Jornadas de Zoología del Uruguay

Verde M, Delgado E, Piñeiro G (1998) *Chasmagnathus granulata* Dana, 1851 (Decapoda, Grapsidae) en la Formación Villa Soriano (Holoceno) del Uruguay. Paper presented at the II Congreso Uruguayo de Geología, Montevideo

Chapter 3
Continental Beds

Abstract The Continental Quaternary sedimentary beds of Uruguay yield useful information for helping to understand the environmental, climatic, and biotic evolution in this region of South America. Most of the units described have a rich paleontological content encompassing wood, fresh-water mollusks, pollen, trace-fossils, and a panoply of vertebrates, with mammals the dominant group (updated tables and selected material on them is included). Lithological units are analyzed in terms of their sedimentary features, the various estimations of their chronology, the depositional environment and fossil content, and its environmental and biogeographic connotation. Economic interest in the units is also brought to light. The Salto, Sopas and Dolores Formations have been selected for detailed discussion, and additional commentaries are provided about the Bellaco, Raigón and Libertad Formations. The sedimentary beds are representative of different time periods, including a fluvial braided system related to a proto Uruguay river in western Uruguay, fluvial contexts with channels and plain-flooded facies along with paleosoils in northern Uruguay, and transitional deposits in southern Uruguay. According to numerical ages, arranged in updated tables, correlation with the Marine Isotope Stage 3 (MIS-3) and MIS-2 is discussed for southern and northern Late Pleistocene beds. In this sense, evidence is analyzed that is based on the fossil content, implying that biogeographic processes—likely related to the climatic conditions—occurred during these time intervals.

Keywords Salto Formation · Sopas Formation · Dolores Formation · Geology · Paleontology · Paleoecology · Radiocarbon · OSL/TL · Trace-fossils

3.1 Geological Features

3.1.1 Salto Formation

According to Veroslavsky and Ubilla (2007), the Salto Formation is a member of the Salto depositional sequence. This sequence is located in northwestern Uruguay

© United Nations University 2016

M. Ubilla and S. Martínez, *Geology and Paleontology of the Quaternary of Uruguay*, SpringerBriefs in Earth System Sciences,
DOI 10.1007/978-3-319-29303-5_3

(Fig. 3.1) and includes braided river deposits, and lacustrine and probably aeolian deposits. The Salto depositional sequence is formed by two cycles (Fig. 3.2); the lower one is represented by most of the Salto Formation, and the upper one includes the uppermost levels of this unit and the gypsum clay bodies of the Bellaco Formation. Goso and Bossi (1966) described the "Areniscas de Salto" of Caorsi and Goñi (1958) as the Salto Formation, but the sediments of this unit were previously reported on by several authors (Frenguelli 1920; Walther 1931; Lambert 1940a, b among others). It crops out discontinuously in western Uruguay (Fig. 3.1), overlies Cretaceous and Late

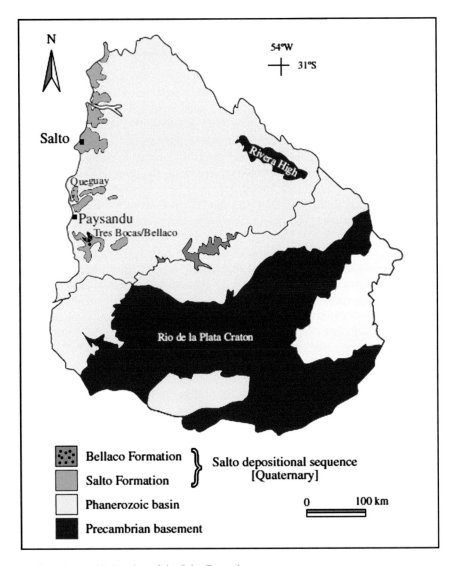

Fig. 3.1 Geographic location of the Salto Formation

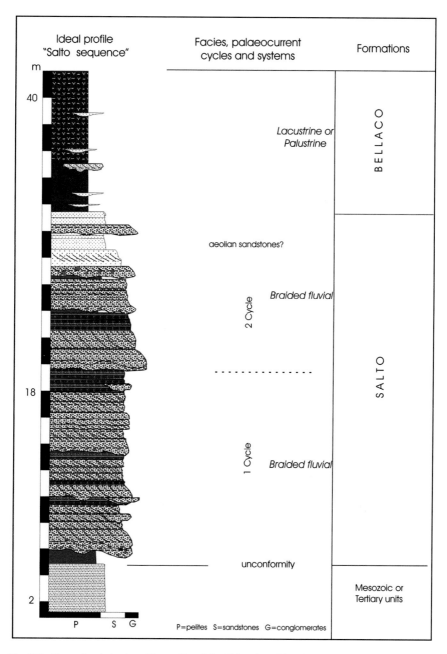

Fig. 3.2 Generalized stratigraphic profile of the Salto depositional sequence including the Salto and the Bellaco Formations (adapted from Veroslavsky and Ubilla 2007)

Paleogene strata, and underlies Pleistocene levels. It has economic value since it is used as an aquifer and is occasionally the substratum for citrus cultivation.

The Salto Formation is more than 35 m thick (including surface and subsurface profiles) and is mostly represented by tabular and lenticular coarse sandstone beds intercalated with sandy conglomerates and fine sandstones and pelites (see details in Veroslavsky and Ubilla 2007; Iriondo and Kröhling 2008) (Figs. 3.2 and 3.3). The lithological features of this unit reveal a sand-dominated braided system, with subordinate pelites and rare conglomerates. The sandstone facies yield trough-cross stratification predominantly on a medium to large scale, horizontal bedding, and massive sandstone; the sandstone includes armored mud balls and are mostly composed of quartz grain, feldspar, opal, and basalt and igneo-metamorphic fragments being the silica the dominant cement (Veroslavsky and Montaño 2004). Ferruginous sandy levels and strongly silicified beds are very common at the top of this unit. The pelites (massive and laminar) are mostly restricted to basal portions of the cycle and have mud cracks. The conglomerates are mostly massive.

Fig. 3.3 **a** Beds of the Salto Formation, **b** coarse to fine sandy beds of the Salto Formation including horizontal and cross-stratification

3.1.2   Sopas Formation

This unit has a patched distribution at river, stream, and creek sides in northern Uruguay (mostly in the Artigas, Salto, Paysandú and Tacuarembó Departments), and usually crops out in a few hundred meters' length and up to 12–15 m in thickness (Fig. 3.4). Antón (1975) described two sedimentary units for northern

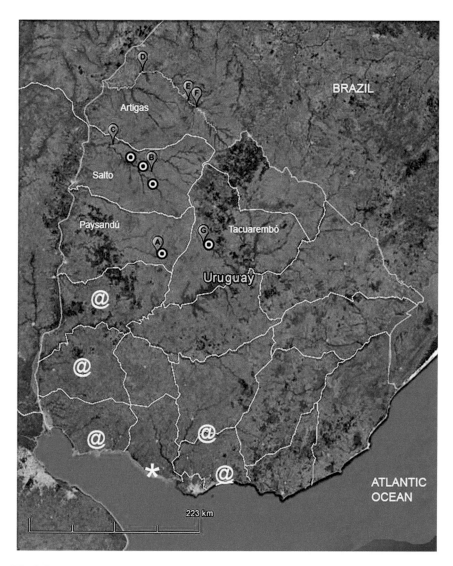

Fig. 3.4   a–g Geographic location of fossiliferous outcrops of the Sopas Formation. *White/black points* indicate geographic location of the ichnofossil *Castrichnus incolumis*, @ areas with outcrops of the Dolores Formation, * geographic location of the Raigón Formation

Uruguay: the Mataojo and Sopas Formations. He characterized the Sopas by massive brownish mudstones with disseminated gravel clasts and carbonates, including in the upper section layers of volcanic ash 30 cm thick. According to him, the Mataojo Formation consists of conglomerates with angular and rounded clasts usually located at the base of stratigraphic profiles. Panario and Gutiérrez (1999) classified the quaternary fluvial deposits of Uruguay as "upper terraces" and "lower terraces" and included the Dolores-Sopas Formation with two members into the lower ones. According to Ubilla and Perea (1999) and Ubilla et al. (2004), the conglomerates of the Mataojo Formation have scarce vertical and horizontal expression, a situation that makes it very difficult to map it so it always appears exposed in relation, and occasionally interbedded, with the brown mudstones of the Sopas Formation. Since it seems inappropriate to separate and define two different lithostratigraphic units, here the Sopas Formation includes those lithofacies that belong to the Mataojo and the Sopas Formation of Antón (1975) (Fig. 3.5). The sediments intermittently overlay Cretaceous basaltic rocks (the Arapey Formation) and Jurassic-Cretaceous aeolian sandstones (Tacuarembó Formation).

Based on several outcrops, Ubilla et al. (2004), and Goso Aguilar and Ubilla (2004) provided a lithofaciological characterization that is briefly synthesized here.

Fig. 3.5 a–d Outcrops of the Sopas Formation showing facies with conglomerates, sand, and mudstones

From the base to the top, the following fining-upwards pattern predominates in sections up to 12 m high (Fig. 3.5):

- Conglomerate with mud matrix (paraconglomerate), reddish and brown color, with pebble- to cobble-size clasts. The composition is basaltic, quartz, aeolian sandstone and chalcedony, both well-rounded and angularly round, although cross-stratification is very common with irregular base contact and channelized geometries that are exposed. The sets are 0.40–1.50 m thick and they can have a rich content of vertebrate and mollusk fossils remains.
- Thin and coarse to fine sandstones with a silt matrix a few centimeters thick, showing ripple and normal graded lamination interbedded with the previous lithology.
- Mudstones and brownish-colored wackestones. Intercalated gravel clasts forming thin bed-sets 3–4 cm thick are also present. This facies yields vertebrates, bivalves, gastropods, and ichnofossils. The wackestone facies exhibit predominantly traction structures. The presence of levels with carbonate concretions, dust and duricrusts that appear mainly at the top of the sections along with rhizo-concretions are very often found.

3.1.3 Dolores Formation

This sedimentary unit described by Goso (1972) is mostly seen in southern Uruguay (the Soriano, Río Negro, Colonia, and Canelones Departments, among others) (Fig. 3.4) with up to 10 m thick outcrops. It overlays the Chuy Formation (Pleistocene), the Fray Bentos Formation (Oligocene) and Precambrian rocks (Preciozzi et al. 1985; Martínez and Ubilla 2004). Beds of the Dolores Formation are studied on coastal cliffs (Canelones and Colonia Departments), rivers, and creek banks (Spoturno et al. 2004; Goso Aguilar 2006). It is characterized by brownish to green/gray siltstones, pelites, and sandy to gravelly pelites and sandstones, with an argillaceous matrix (Preciozzi et al. 1985; Bossi and Navarro 1991) (Fig. 3.6). Carbonate, massive silty-clay sediments, parallel lamination, incised lobes, and tabular deposition were also described (Goso Aguilar 2006). The newest soils of southern Uruguay have been developed from this unit. It is different from the Libertad Formation due to some of its geomorphological features (Martínez and Ubilla 2004).

3.1.4 Other Units

The Raigón Formation, described by Goso and Bossi (1966) (similar to the San José Formation of Francis and Mones 1965), outcrops in southwestern Uruguay's coastal cliffs of 5 m thick on the surface, but up to 50 m in the subsurface (Fig. 3.4).

Fig. 3.6 a–d Outcrops of the Dolores Formation at the Santa Lucía Basin (southern Uruguay)

It is mostly characterized by fining upward to coarse sandy levels, including parallel and cross-bedding stratification, with intercalated massive silty-clayey greenish beds (Fig. 3.7) (Perea and Martínez 2004; Tófalo et al. 2009). Bossi et al. (2009) described two members of this unit, from the base to the top—the San José Member and the San Bautista Member. According to these authors, sandstones and conglomerates, along with fine clayish sandstones, dominate the first; the second is represented by loess that constitutes paleosoil evolution. Similar to the Salto strata, the Raigón sedimentary bed is also an important aquifer strongly related to agro-industrial activities.

Goso and Bossi (1966) described the Libertad Formation in southern Uruguay which, according to Preciozzi et al. (1985), is up to 30 m high (with the surface and subsurface). It lies unconformable on Paleozoic and tertiary rocks and under Late Quaternary beds or recent soils (Tófalo et al. 2009). It is mainly characterized by brown mudstones with scattered coarse sand, loess and calcium carbonate. Paleosoils, tabular geometry and coarse stratification were observed, but the absence of internal sedimentary structures dominates (Bossi et al. 2009; Tófalo et al. 2009). Geomorphological studies envisage at least two depositional episodes (Libertad I and Libertad II) related to climatic changes (Panario and Gutiérrez 1999). This unit was also interpreted as the result of the weathering of various types of rocks, especially those of the San Bautista member of the Raigon Formation (Bossi et al. 2009).

Fig. 3.7 a–b Outcrops of the Raigon Formation (*arrow* indicates upper limit), **c** cross-stratification in the Raigón Formation

3.2 Numerical Ages

3.2.1 Salto Formation

There are various opinions on the age of the Salto Formation; they are mostly based on stratigraphic evidence or correlation with climatic processes. Stratigraphic

Table 3.1 Thermoluminescence (TL) and Optically Stimulated Luminescence (OSL) ages of the Salto and the Raigón Formations

ID lab Formation	S	Sample location	TL/OSL Age (yr)
SALTO Formation			
–	Fs	Near Salto	88,370 ± 35,680
LVD-948	Fs	Salto City	986,000 ± 100,000
LVD-949	Fs	Salto dep.	830,000 ± 95,000
RAIGÓN Formation			
LVD-1450	Ms	Arazati, SJ	218,000 ± 26,000
LVD-1451	Ms	Ordeig, SJ	100,000 ± 12,000
UIC-3457	Fs	Ordeig, SJ	373,765 ± 28,455**
UIC-3334	Fs	Ordeig, SJ	>180,000**
UIC-3335	Ms	Arazatí, SJ	>390,000**
UIC-3337	Cs	Arazatí, SJ	>222,000**
UIC-3456	Ms	Arazatí, SJ.	>230,000**
UIC-3452	Fs	Arazati, SJ	>154,000**

Iriondo and Kröhling (2003), Veroslavsky and Ubilla (2007), Ubilla et al. (2009), **this paper
S sample, Fs fine sand, Ms medium sand, Cs coarse sand, SJ San José

relationships, as previously mentioned, indicate an age not older than the Late Oligocene and not younger than the Late Pleistocene. This unit was considered Late Miocene, Pliocene, and Pleistocene by various authors (see details in Table 2 in Veroslavsky and Ubilla 2007; Panario et al. 2014). There are numerical ages based on Optically Stimulated Luminescence and Thermoluminescence (OSL/TL) methods that reinforce the estimation of a Pleistocene age suggested by a few authors (see Goso 1972; Antón 1975; Goso and Bossi 1966). Two ages were proposed that were based on sandy friable samples from two localities taken from levels of the lower cycle of the Salto sequence (Veroslavsky and Ubilla 2007): 986 ± 100 ky and 830 ± 95 ky (Table 3.1). In this sense, an Early-Middle Pleistocene age is likely to be presumed. In addition, a TL age of 88.370 ± 35.680 years is assumed from the upper portion of the unit (Salto Department), related to the Late Pleistocene (Iriondo and Kröhling 2008).

3.2.2   Sopas Formation

According to the paleontological content (vide infra), a Late Pleistocene age was proposed for this unit, and a biostratigraphic correlation with the Lujanian stage/age (Late Pleistocene/Early Holocene) of the Buenos Aires Province was also postulated (Ubilla and Perea 1999; Ubilla et al. 2004). In the last few years, the number of numerical ages available for this unit successfully increased (Tables 3.2 and 3.3). Several conventional and AMS radiocarbon and OSL/TL ages were produced from different outcrops using samples of wood, fresh-water mollusk shells, mammal

Table 3.2 ^{14}C conventional and AMS ages from the Sopas Formation

ID Lab.	Taxon sample	M	Sample location	^{14}C age BP (Cal BP***)
LP-594	Wood indet	Wood	RCA	12,100 ± 140 (13,550–14,373)
AA104912*	*Cyanocyclas* sp.	Shell	TARN	12,502 ± 55 (14,234–15,001)**
AA99843*	Wood indet	Wood	RCA	13,869 ± 54 (16,473–16,983)**
AA104915*	*Pomacea* sp.	Shell	MT	33,560 ± 700 (36,089–39,426)**
AA101329*	*Pomacea* sp.	Shell	MT	35,530 ± 680 (38,659–41,421)**
AA104914*	*Pomacea* sp.	Shell	MT	37,070 ± 810 (39,940–42,665)**
AA104913*	*Pomacea* sp.	Shell	MT	38,300 ± 940 (40,865–43,932)**
AA104911*	*Cyanocyclas* sp.	Shell	MT	39,900 ± 1,100 (42,025–45,389)**
AA101328*	*Diplodon* sp.	Shell	MT	>45,200**
URU-0032	*D. peraeformis*	Shell	MT	>45,000
URU-0031	*D. peraeformis*	Shell	MT	>45,000
URU-0053	*Prosopis nigra*	Wood	CSS	>45,000
LP-490	*Prosopis* sp.	Wood	RCA	>43,000
URU-0036	*Prosopis* sp.	Wood	RCA	>45,000

Ubilla and Perea (1999), Ubilla et al. (2004), **Ubilla et al. accepted (2015)
*AMS ones
***95.4 (2 sigma), *M* material, *RCA* Río Cuareim (Artigas), *TARN* Tres Arboles creek (Río Negro), *MT* Malo creek, (Tacuarembó), *CSS* Cañada Sarandí (Salto)

bones and teeth, and sediment. Unfortunately, bones and teeth provided little information (Ubilla 2001; Martínez and Ubilla 2004).

Some radiocarbon ages are interpreted as minimum ages, but many finite ages were also produced (Table 3.2). A set of radiocarbon ages range from 33,560 ± 700 yrs BP (cal 36,089–39,426 yrs) to 39,900 ± 1,100 (cal 42,025–45,389 yrs), which correlate with the MIS-3. There are some outcrops with radiocarbon ages ranging from 12,100 ± 140 yrs. BP (cal 13,550–14,373 yrs) to 13,869 ± 54 yrs. BP (cal 16,473–16,983 yrs) that belong to younger facies of this unit. In any case, these ages corroborate the Late Pleistocene age of the unit based on the paleontological content.

A set of OSL/TL ages was produced from samples taken mostly from fossiliferous outcrops (Table 3.3) (Ubilla et al. accepted 2015). In particular, those ranging from 27,400 ± 3,300 to 71,400 ± 11,000 yrs also support a relationship with the Marine Isotope Stage 3 (MIS-3). Most ages fall in the 50–25 ka time interval and it is more frequently represented by the 45–28 ka time interval. Some of the older ages proposed should be taken with warning, because some have stratigraphic inversion or totally divert from this general pattern, becoming harder to explain.

Table 3.3 Optically Stimulated Luminescence (OSL) and Thermoluminescence (TL) ages of the Sopas Formation

ID Lab	S	Sample location	OSL/TL Age (yr)
UIC-3455	ms	ACr, Salto	14,485 ± 1,240**
LVD-1449	ms	Cr, Artigas	27,400 ± 3,300**
LVD-2657	Ss	Cr, Artigas	30,300 ± 3,700***
LVD-2660	sS	Sc, Salto	30,600 ± 5,400***
UIC-3458	ms	Mc, Tacuarembó	32,850 ± 1,990**
UIC-3451	ms	Mc, Tacuarembó	32,995 ± 1,930**
UIC-3332	ms	Mc, Tacuarembó	34,405 ± 2,240**
LVD-2655	Ss	Cr, Artigas	36,100 ± 6,200***
LVD-2661	ms	ACr, Salto	36,900 ± 6,500***
LVD-647	sSC	Sc, Salto	43,500 ± 3,600
LVD-646	sS	Mc, Tacuarembó	58,300 ± 7,400
LVD-2658	Ss	Cr, Artigas	71,400 ± 11,000***
LVD-1241	ms	Cr, Artigas	96,000 ± 11,000**
LVD-859	sS	Cr, Artigas	180,000 ± 20,000
LVD-857	s	Mc, Tacuarembó	200,000 ± 25,000
LVD-2659	Ss	Cr, Artigas	248,000 ± 26,000***
LVD-1242	sS	Mc, Tacuarembó	314,000 ± 39,300**
LVD-858	fs	Cr, Artigas	360,000 ± 40,000

Ubilla (2004), Ubilla et al. (2004), Martínez and Ubilla (2004), ***Prosul (2009–2011), **Ubilla et al. accepted (2015)

S sample, *ms* medium sand, *Ss* silty sand, *sS* sandy silt, *SsC* sandy silt crotovina, *s* silt, *fs* fine sand, *ACr* Arapey Chico River, *Cr* Cuareim River, *Sc* Sopas Creek, *Mc* Malo Creek

3.2.3 Dolores Formation

In the last few years, many numerical ages based on radiocarbon and OSL/TL methods were produced (the Rio Negro Department and several outcrops of the Santa Lucía Basin in southern Uruguay). According to this information, the last 30–10 ky lapse of time is represented (Tables 3.4 and 3.5). A set of radiocarbon ages based on organic soil, wood, and mammal teeth ranges from ^{14}C age of 22,450 ± 400 yrs BP (cal BP 25,934–27,436) to 10,140 ± 50 yrs BP (cal BP 11,857–11,960) (Table 3.4). The OSL/TL methods produced ages of 32,230 ± 2,640 yrs to 10,570 ± 990 yrs (Santa Lucía River Basin) (Table 3.5), which are stratigraphically consistent with regard to radiocarbon ages.

In previous studies, the Dolores Formation was interpreted as a Late Pleistocene unit based mostly on stratigraphic relationships and mammalian content (Preciozzi et al. 1985; Ubilla and Perea 1999; Martínez and Ubilla 2004), but the numerical dating indicates a Late Pleistocene to Early Holocene age.

Table 3.4 ¹⁴C conventional and AMS ages from the Dolores Formation

Lab ID	Taxón sample	Material	SL	¹⁴C age BP (Cal BP age***)
Beta301006*	–	Organic soil	Vc	10,140 ± 50 (11,857–11,960)
LP-1110	*Salix humboldtiana*	wood	SLrPC	10,480 ± 105 (11,982–12,647)
LP1143	–	wood	SLrPP	10,500 ± 110 (11,992–12,656)
LP1268	*Prosopis* sp.	wood	Vc	11,090 ± 110 (12,722–13,090)
LP1283	–	wood	SLrB	11,150 ± 120 (12,725–13,169)
AA91726*	*Equus* sp.	enamel	Pc	21,530 ± 140 (25,525–26,037)**
AA99845*	Deer	enamel	Pc	22,450 ± 400 (25,934–27,436)

Ubilla (1999), Martínez and Ubilla (2004), Meneghin (2011), Ubilla and Rinderknecht (2014a), **this work
*AMS ones
SL sample location, ***2 Sigma, *Vc* Vejigas Creek, *SLrPC* Santa Lucía River (Paso Cuello), *SLrPP* Santa Lucía River (Paso Pache), *SLrB* Santa Lucía River (Barrancas), *Pc*: Pilatos Creek

Table 3.5 Optically Stimulated Luminescence (OSL) and Thermoluminescence (TL) ages of the Dolores Formation

ID Lab	S	Sample location	OSL/TL Age (yr)
UIC-3052	fs	Vc Canelones	10,570 ± 990
UIC-3040	ms	EChc Colonia	15,730 ± 925
UIC-3039	ss	ECac Colonia	16,070 ± 930
UIC-3053	fs	Pc Canelones	23,785 ± 2,990
UIC-2822	fs	Pc Canelones	30,855 ± 2,370
UIC-2826	fs	Ac Canelones	31,160 ± 2,285
UIC-3302	ss	Ac Canelones	32,230 ± 2,640

Ubilla et al. (2013), Corona et al. (2013), Ubilla and Rinderknecht (2014a)
S sample, *fs* fine sand, *ms* medium sand, *ss* silty sand, *Vc* Vejigas Creek, *EChc* El Chileno Creek, *ECac* El Caño Creek, *Pc* Pilatos Creek, *Ac* Aparicio Creek

3.2.4 Other Units

The Raigón Formation, usually consistent with the Salto Formation, has been considered a Pliocene unit, but there is some evidence from upper levels that suggest also a Pleistocene age. Several OSL ages were produced from medium and upper beds using sandy samples (Table 3.1). Most are minimum ages, but there is one finite age of 373,7 ± 28,4 ky, indicating Medium Pleistocene, along with an OSL age of 218 ± 26 ky and 100 ± 12 ky from upper beds of the latest Medium and

Late Pleistocene. Bossi et al. (2009) proposed a Late Pliocene to Middle Pleistocene age, and Tófalo et al. (2009) a Late Pliocene to Early Pleistocene.

In northern Uruguay (Cuareim and Uruguay Rivers), there are sedimentary beds that are particularly interesting for archaeological studies (MEC 1989; Castiñeira et al. 2010; Suárez 2011; López Mazz 2013). It must be noted that most of the authors did not mention the sedimentary context of the Sopas Formation, except for Castiñeira et al. (2010). A large number of ^{14}C ages were produced, ranging from approximately 11 to 8,5 ky BP (Suárez and Lopez 2003; Castiñeira et al. 2010; Suárez and Santos 2010; Suárez 2011; López Mazz 2013, and references therein) (Table 3.6). The mammals *Equus* sp. and *Glyptodon* sp. were reported in association with lithic material in a 9 ky BP level (Suárez and Santos 2010). A calibration of the 11,200 ± 500 year BP age (MEC 1989) provides a 2 sigma cal BP 11,600–14,176 yrs, a roughly similar age for the Sopas Formation at the Cuareim River in northern Uruguay (cal 13,550–14,373 yrs).

The Libertad Formation has been considered by stratigraphic relationships to be a Lower to Middle Pleistocene unit (Martínez and Ubilla 2004). Recently, its age was considered to be around 20 ky, based on U-Th in *Macrauchenia* (Cid et al. 2014) and 17 ky for *Stegomastodon* (Gutiérrez et al. 2005) of a bone-bed in southern Uruguay (see below). Several studies of southern and eastern Uruguay (mostly related to archaeological excavations) provided many radiocarbon ages from different localities, ranging from 30 to 3 ky (Table 3.7) (Meneghin 2004, 2006, 2015; López Mazz 2013; Fariña et al. 2013, among others).

Table 3.6 Selected ^{14}C conventional and AMS ages from archaeological sites in northern Uruguay

ID Lab.	Material	SL	^{14}C age BP	Cal BP
URU 0079	charcoal	UrS	4,020 ± 70	–
URU 248	charcoal	PPA	8,570 ± 150	9,370–9,680
S7n	charcoal	CD3NS	9,300 ± 150	8,719–8,326
Rt 1445	charcoal	PPA	9,890 ± 75	9,450–9,261
UCLAMS21641*	charcoal	PPA	9,585 ± 25	10,960 > 11,000
UCLAMS28692*	charcoal	PPNA	10,465 ± 30	10,579–10,285
Kn2531	charcoal	IdTS	10,429 ± 90	10,571–10,194
UCLAMS21631*	charcoal	PPA	10,930 ± 20	10,919–10,763
GIF4412	charcoal	Ida	11,200 ± 500	11,600–14,176

Guidón (1989), Hilbert (1991), Austral (1995), Castiñeira et al. (2010), Suárez and Santos (2010), López Mazz (2013), Suárez 2009 in López Mazz (2013)
*AMS ones
SL sample location, *UrS* Uruguay river (Salto), *PPA* Pay Paso (Artigas), *CD3 NS* CalpicaDO3N (Salto), *PPNA* Pay Paso Norte, *IdTS* Isla del Tigre (Salto), *Ida* Isla de Arriba

Table 3.7 Selected ^{14}C conventional and AMS ages from archaeological sites in southern and eastern Uruguay

ID Lab.	Material	Locality	^{14}C age BP	Cal BP
Beta286135*	charcoal	UM	2,900 ± 40	3,170–2,930
URU-0515	charcoal	LIR	7,100 ± 160	6,202–5,789
CURL6078*	charcoal	LIR	8,510 ± 40	7,583–7,543
Beta165076*	charcoal	UM	10,690 ± 60	12,620–12,960
Beta380727*	charcoal	UM	10,800 ± 30	12,725–12,695
Beta211938*	charcoal	UM	11,690 ± 80	13,430–14,020
Beta395639*	charcoal	UM	12,000 ± 40	13,835–13,735
URU-0496	Bone *Lestodon*	VcC	27,000 ± 450	29,696 ± 871
URU-0493	Bone *Lestodon*	VcC	30,100 ± 600	32,886 ± 1,446

Meneghin (2004, 2015), López Mazz et al. (2009), Fariña et al. (2013)
*AMS ones
UM Urupez (Maldonado), *LIR* Los Indios (Rocha), *VcC* Vizcaíno Creek (Canelones)

3.3 Paleontological Context

3.3.1 Salto Formation

The fossil content of the Salto Formation has been scarce until now. It is represented by the silicified wood of small to medium trees referred to as Leguminosae and Caesalpinoideae without chronological information (Aznárez 1945) (Fig. 3.8). In general, there are preserved fragments or large portions of trunks that are likely parautochthonus revealing local transport. However, they have been referred trunks in life position (Iriondo and Kröheling 2008). The presence of the gastropod *Eoborus berroi* in this unit (Klappenbach and Olazarri 1986) was afterwards rejected and interpreted as reworked material from an older unit (Martinez et al. 1997).

3.3.2 Sopas Formation

This is a very fossiliferous sedimentary unit that includes trace-fossils (coprolites, nests, and caves) and body fossils (wood, fresh-water mollusk shells, and vertebrates). This fossil assemblage provides useful information for interpreting the climatic and environmental conditions involved.

Trace-fossils are represented by some burrow-like structures. The likely trace-producer is the now-extinct rodent *Microcavia criolloensis* (Ubilla et al. 1999) that was found to be associated with the burrows (Ubilla 2008); there are also structures interpreted as large paleocaves (Sopas Creek) (Fig. 3.9). The coprolites (Cuareim River) are related to medium to large predators (Verde and Ubilla 2002), and a canid origin was proposed (Chimento and Rey 2008). In fact, the

Fig. 3.8 Silicified wood of
the Salto Formation

hypercarnivorous canids, such as *Protocyon* or *Dusicyon avus* (Prevosti et al. 2009)
can be considered as possible producers. A very unique and abundant type of
preservation is represented by *Castrichnus incolumis* and *Taenidium serpentinum*
that were described by Verde et al. (2007) (Fig. 3.9), which was interpreted as
earthworm estivation chambers produced in paleosoils (Sopas and Arerunguá
Creeks, Salto; Malo Creek, Tacuarembó; Queguay River, Paysandú).

The wood remains provided limited information and some were determined to
be *Prosopis* (Inda and del Puerto 2002; Ubilla et al. 2004; Martínez and Ubilla
2004) (Fig. 3.10). Fresh-water bivalves, gastropods, and a few terrestrial snails were
described in several outcrops (Martínez and Rojas 2004). The bivalves are frequently found with articulated valves, and the gastropods are usually complete
(Fig. 3.11).

The vertebrates of the Sopas Formation include a few Teleostei indet., reptiles,
some birds, and numerous mammals (Tables 3.8 and 3.9) (Ubilla et al. 2004, 2011;
Ubilla et al. accepted 2015). Large extinct terrestrial tortoises, the flightless bird
Rhea, the seriema *Cariama*, and the Magellan-goose *Chloephaga* are among extant
birds recorded in the Sopas Formation (Tambussi et al. 2005). Mammals are the
dominant group (25 families in 9 orders, encompassing more than 50 species) and
many extinct taxa and extinct species of extant genera are recorded. There are some
taxa not represented in the present-day communities of Uruguay, but currently live
in other areas of South America, such as some rodents, peccaries, and tapirs

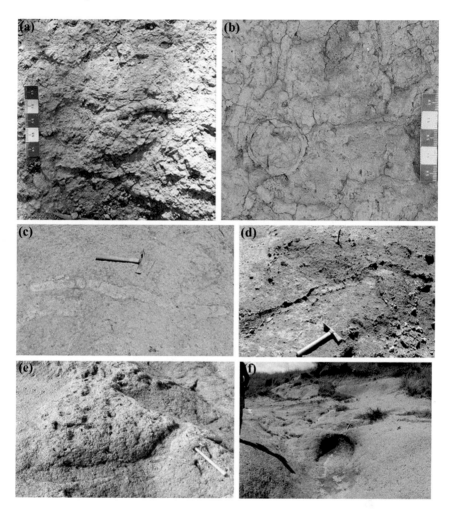

Fig. 3.9 a–b *Castrichnus incolumis* and *Taenidium serpentinum*, **c–e** rodent burrows, **f** paleocave

(Fig. 3.12). It shows local extinctions and shifting ranges. Ungulates (especially deer) and rodents are the groups most frequently represented in the fossil assemblage. Some very large to small herbivores, such as *Toxodon, Macrauchenia,* the horses *Equus neogaeus* and *Hippidion principale,* the ground sloth *Glossotherium, Neolicaphrium, Lama, Microcavia,* some omnivores, like the bear *Arctotherium* and large- to medium-sized predators like the jaguar *Panthera onca,* the mountain-lion *Puma,* and the extinct canid *Protocyon,* among others, were recorded (Table 3.9) (Fig. 3.13).

Fig. 3.10 a–b Wood remains of the Dolores Formation, **c** wood of the Sopas Formation

3.3.3 Dolores Formation

This unit yields pollen, a few fresh-water mollusks, a few turtles, and an important number of mammals that provide tools to understand environment and climatic conditions (Ubilla et al. 2009, 2011; De Oliveira et al. 2011; López Romanelli 2012). Preliminary pollen studies from facies of 11 to 10 ky reveal a dominance of

Fig. 3.11 a Gastropods
(*Pomacea* sp.) of the Sopas
Formation, **b** Articulated
bivalves (*Diplodon* sp.) in the
Sopas Formation

herbs, since trees and shrubs are scarce (De Oliveira et al. 2011). Typically extinct
South American Pleistocene mammals are present; these include megafaunal representatives. *Macrauchenia patachonica*, *Toxodon* cf. *T. platensis*, the ground
sloths *Glossotherium robustum* and *Catonyx cuvieri*, the glyptodonts *Glyptodon
clavipes*, *Doedicurus clavicaudatus* and *Panochthus* cf. *P. tuberculatus*, armadillos
as *Pampatherium typum*, the camelid *Hemiauchenia* sp., the horses *Equus neogeus*
and *Hippidion* sp., gomphotheriid *Stegomastodon*, deer such as *Morenelaphus* sp.,
the sabre-tooth *Smilodon populator* and medium-size bear like *Arctotherium tarijense*, among others (Fig. 3.14). Some extinct large mammals (*Glyptodon*,
Morenelaphus, large camelids) were recorded in beds with Early Holocene radio-
carbonic ages. This information suggests that these kinds of animals could have
survived until at least the early phases of Holocene at these latitudes.

Table 3.8 Updated list of non-mammal vertebrates for the Sopas Formation of northern Uruguay

Teleostei Paracanthopterygii/Acanthopterygii indet.
Testudines Family Testudinidae 　*Chelonoides sp.*
Squamata Family Teiidae 　*Tupinambis* cf. *T. teguixin*
Aves Family Rheidae 　*Rhea* sp. Brisson
Family Anatidae 　*Chloephaga picta*
Family Cariamidae 　*Cariama cristata*
Family Psitaciidae 　*Cyanoliseus patagonus*
Family Furnariidae 　cf. *Pseudoseisuropsis* sp.

Ubilla et al. (2004), Tambussi et al. (2005, 2009) and this work

Table 3.9 Updated list of mammals for the Sopas Formation of northern Uruguay

Order Didelphimorphia Family **Didelphidae** 　cf. *Didelphis* sp.	**Order Litopterna** Family **Macraucheniidae** 　*Macrauchenia patachonica* Family **Proterotheriidae** 　*Neolicaphrium recens* 　*N.* cf. *N. recens*
Order Xenarthra Family **Dasypodidae** 　*Dasypus* aff. *D. novemcinctus* 　*Propraopus* sp. Family **Pampatheriidae** 　*Pampatherium typum* 　*Pampatherium humboldti* Family **Glyptodontidae** 　*Glyptodon clavipes* 　cf. *Hoplophorus* 　*Neuryurus rudis* 　*Panochthus tuberculatus* Family **Megatheriidae** 　*Megatherium americanum* Family **Nothrotheriidae** 　*Nothrotherium* cf. *N. maquinense* Family **Mylodontidae** 　*Glossotherium robustum* 　*Lestodon armatus* 　*Catonyx cuvieri*	**Order Notoungulata** Family **Toxodontidae** 　*Toxodon* cf. *T. platensis* **Order Proboscidea** Family **Gomphotheriidae** indet. **Order Perissodactyla** Family **Tapiridae** 　*Tapirus terrestris* 　*Tapirus* sp. Family **Equidae** 　*Equus (Amerhippus) neogeus* 　*Hippidion principale*

(continued)

Table 3.9 (continued)

Order Carnivora	Order Artiodactyla
Family **Canidae**	Family **Tayassuidae**
Lycalopex gymnocercus	*Tayassu pecari*
Dusicyon avus	*Catagonus* wagneri
Protocyon troglodytes	*Catagonus stenocephalus*
Family **Felidae**	Family **Cervidae**
Felis concolor	*Antifer ultra*
Panthera cf. *P.onca*	*Ozotoceros* aff. *O. bezoarticus*
Smilodon populator	*Morenelaphus brachyceros*
Family **Mustelidae**	*Morenelaphus lujanensis*
Lontra longicaudis	*Paraceros fragilis*
Family **Ursidae**	*Mazama* sp.
Arctotherium aff. *A. bonariense*	Family **Camelidae**
	Hemiauchenia paradoxa
	Lama guanicoe
	Palaeolama major
	Vicugna vicugna
Order Rodentia	
Family **Cricetidae**	
Reithrodon sp.	
cf. *Wilfredomys oenax*	
Lundomys molitor	
Family **Erethizontidae**	
Coendou magnus	
Family **Echimyidae**	
Myocastor coypus	
Family **Chinchillidae**	
Lagostomus sp.	
Family **Caviidae**	
Cavia sp.	
Galea sp.	
Microcavia criolloensis	
Dolichotis sp.	
Hydrochoerus hydrochaeris	
Neochoerus cf. *N. aesopi*	

Ubilla et al. (2004, 2009, 2011) and references therein, Perea (2007), Prevosti et al. (2009), Scherer (2009), Gasparini et al. (2009, 2013), Corona et al. (2012), and this work

Also recorded are some mammals that are now extinct in this area but still extant in other regions of South America at the generic or specific level (the rodents *Microcavia, Galea, Dolichotis* cf. *D. patagonum, Lagostomus maximus,* the armadillo *Chaetophractus villosus,* the camelid *Vicugna* sp.) (Fig. 3.15) (Rego et al. 2007; Ubilla 2008; Ubilla et al. 2009, 2011; Corona et al. 2013; Ubilla and Rinderknecht 2014a, b). They illustrate shifting ranges or local extinctions. Most of the tropical to subtropical mammals found in the Sopas Formation (northern Uruguay) are absent.

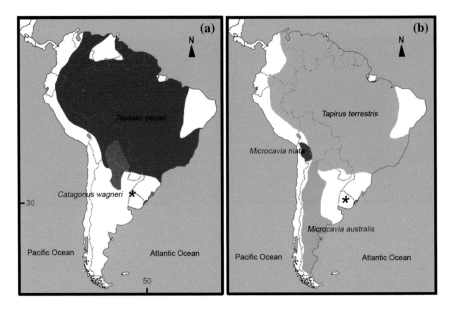

Fig. 3.12 **a** Current distribution of *Catagonus wagneri, Tayassu pecari* and their fossil record in Late Pleistocene of Uruguay (Sopas Formation), **b** Current distribution of *Tapirus terrestris*, *Microcavia* spp. and their fossil record in the Late Pleistocene of Uruguay (Sopas Formation). (Quintana 1996; Ubilla et al. 2004; Pardiñas and Ojeda 2008; Gasparini et al. 2013; Keuroghlian et al. 2013; Naveda et al. 2015)

3.3.4 Other Units

The Raigón Formation yields little yet interesting fossil content, mostly represented by vertebrates (Perea and Martínez 2004; Perea et al. 2013). Some ground-sloths and glyptodonts, toxodonts, large extinct phorusrhacids birds along with medium to giant rodents such as dinomyids (Tambussi et al. 1999; Rinderknecht and Blanco 2008; Perea et al. 2013) are among the most significant fossils (Fig. 3.16). Some mammals suggest a Pliocene to Medium Pleistocene age (Perea et al. 2013).

Probably belonging to the Libertad or Dolores Formation were some frogs that were described as *Ceratophrys* and *Leptodactylus*, a few birds, such as *Colaptes* and *Pseudosesiuropsis,* along with some extinct large mammals (Ubilla et al. 2011 and references therein). Many fossils were doubtfully assigned to the Libertad Formation (Ubilla and Perea 1999). However, a Late Pleistocene bone bed from southern Uruguay, including many taxa of vertebrates, dominated by mammals, is confidently referred to the Libertad Formation. Disarticulated and fractured bones are dominant in the assemblage that originated under a non-channelized mudflow. Glyptodonts, proboscideans, toxodonts, a large bear, and deer are the most representative taxa included in the bone bed (Corona et al. 2012).

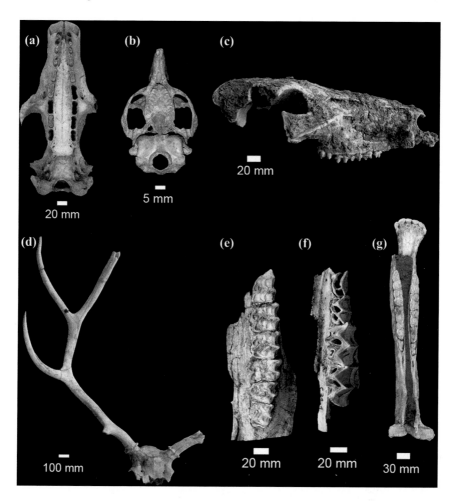

Fig. 3.13 Selected mammals of the Sopas Formation: **a** skull of the xenarthran *Pampatherium humboldti* (palatal view), **b** skull of the rodent *Microcavia criolloensis* (palatal view), **c** skull of the pecari *Catagonus wagneri* (lateral view), **d** skull and antler of the deer *Paraceros fragilis*, **e** upper dentition of the horse *Hippidion principale*, **f** upper dentition of the litoptern *Macrauchenia patachonica*, **g** mandibles of the horse *Equus neogeus*

There is another bone bed in southern Uruguay without reference to any sedimentary unit, including a variety of extinct giant mammals of 30 ky in age, where the ground-sloth *Lestodon* is predominant. Marks in some bones were interpreted as a result of human actions (Fariña et al. 2013), but the evidence is controversial (Suárez et al. 2014).

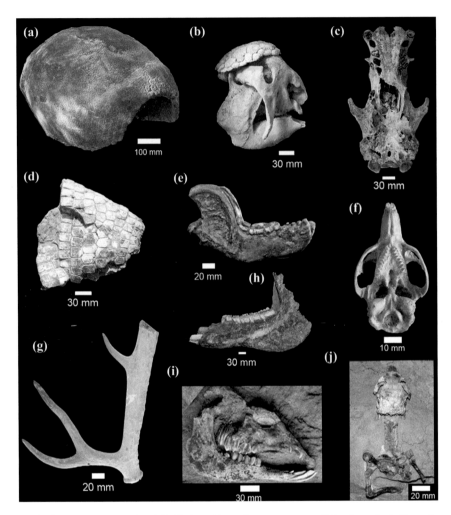

Fig. 3.14 Selected mammals of the Dolores Formation: **a** carapace of the glyptodont *Glyptodon clavipes*, **b** skull with dorsal shield (lateral view) of the glyptodont *Panochthus tuberculatus*, **c** skull (palatal view) of the ground sloth *Glossotherium robustum*, **d** anterior shoulder with fixed scutes of the xenarthran *Pampatherium typum*, **e** lateral view of the mandible of *Arctotherium tarijense*, **f** skull (palatal view) of the rodent *Galea ortodonta*, **g** antler of *Morenelaphus* sp., **h** mandibles of *Toxodon platensis*, **i** skull and mandible (lateral view) articulated of a juvenile of the camelid *Hemiauchenia*, **j** an almost complete and articulated skeleton of the rodent *Microcavia criolloensis*

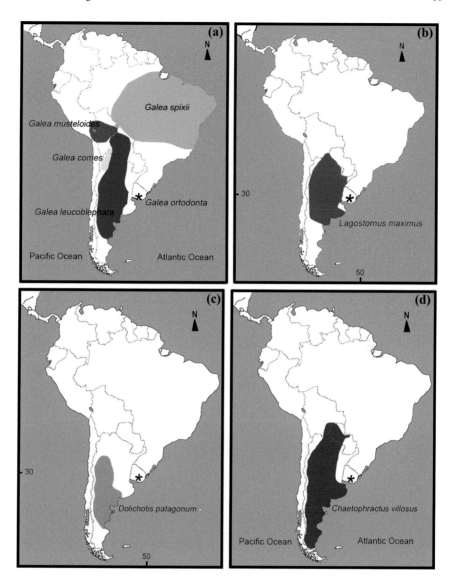

Fig. 3.15 **a** Current distribution of the rodent *Galea* spp. and the extinct *G. ortodonta* in Late Pleistocene of Uruguay (Dolores Formation), **b** current distribution of the rodent *Lagostomus maximus* and its record in the Late Pleistocene of Uruguay (Dolores Formation), **c** current distribution of the rodent *Dolichotis patagonum* and its record in the Late Pleistocene of Uruguay (Dolores Formation), **d** current distribution of *Chaetophractus villosus* and its Late Pleistocene record in Uruguay (Dolores Formation). (Abba et al. 2014; Ubilla and Rinderknecht 2014a, b; Patton et al. 2015)

Fig. 3.16 a Tibiotarsus of a
large carnassial bird
(Phorusrhacine) of the Raigón
Formation compared with the
extant large flightless bird
Rhea sp. (**b**)

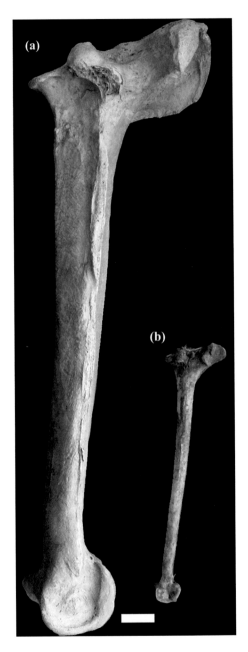

3.4 Environment and Climatic Scenarios

3.4.1 Salto Formation

According to the interpretation of the lithological features of the Salto Formation, the depositional environment was a fluvial braided system. It is considered to be related to a proto Uruguay River, actually a member of the Del Plata Basin (Veroslavsky and Ubilla 2007; Iriondo and Kröhling 2008; Panario et al. 2014). It primarily represents bedload deposits with low sinuosity and highly mobile broad and shallow channels. The armored mud balls well illustrate an erosive and auto-destructive behavior of the system. There is a tendency to assume that the Salto Formation was generated under arid or semi-arid climatic conditions (Bossi and Navarro 1991; Panario and Gutiérrez 1999); sedimentological, petrological and mineralogical information concur with this interpretation. In addition, if the gypsum clay beds of the Bellaco Formation are genetically associated with the Salto Formation, it strongly reinforces the evidence of an arid climate (Veroslavsky and Ubilla 2007). On the other hand, the OSL ages produced from the lower cycle indicate a correlation with some warm episodes of the Middle Pleistocene (Bradley 2015) such those corresponding to the OIS 25, 21, and 19.

Since both units, the Salto and the Bellaco Formations, today occupy high and low topographic locations in western Uruguay, there was invoked localized uplift by some authors (Bossi and Ferrando 2001 and references therein).

3.4.2 Sopas Formation

The sedimentary beds of this unit originated predominantly under fluvial contexts including channel and plain-flooded facies along with paleosoils observed in some localities (Ubilla et al. 2004; Goso Aguilar and Ubilla 2004). A variety of habitats can be assumed, based on the paleontological content (Ubilla et al. 2004). Fresh-water mollusks, the winter migratory bird *Chloephaga*, and some mammals (tapirs, marsh rice rats, capybaras, and river otters), represent lacustrine and fluvial environments. The avian and mammalian assemblage include taxa not only related to fluvial environment contexts, but also with open to semi-forested and forested areas (*Rhea*, *Cariama*, the horses *E. neogeus* and *H. principale*, the deer *Mazama*, and the rodent coendou, among others). A $\delta^{13}C$ isotope data for some ungulates (*Hippidion* cf. *H. principale*, *Equus neogeus*, deer, and a large camelid) indicate predominantly browser to mixed feeding habits likely related to semi-open environments (Morosi and Ubilla 2014). Some mammals (some rodents, peccaries, and camelids) also indicate arid to semi-arid environments, and the earthworm estivation chambers of *Castrichnus* are related to paleosoil development. According to the fossil content, the influence of the Last Interglacial or the Last Interstadial (MIS-3) was brought to mind (Ubilla et al. 2004). Extant representatives of mammalian taxa suggest a relationship with a benign climatic condition (inhabitants of tropical to temperate areas in South America). However, arid to semi-arid indicators were also found and are widespread today in mid- to high latitudes of South America. There is also evidence in favor of seasonality and perhaps droughts based on estivation chambers and winter migratory birds (Tambussi et al. 2005; Verde et al. 2007; Genise et al. 2013). These environmental and climatic conditions could have been developed by the influence of the MIS-3 climatic context that implied millennial climatic changes. The presence of tropical to subtropical taxa could be explained as survivors in environment refuges during the MIS-3 at this latitude due to the presence of perennial rivers, riparian forests and semi-forested areas (Ubilla et al. accepted 2015).

3.4.3 Dolores Formation

The sedimentary beds of this unit belong to continental environments (including gravity flows and reworked aeolian deposits) associated with a cold and arid to semi-arid climate (Martínez and Ubilla 2004; Goso Aguilar 2006). Open to semi-open environments and a predominance of grasslands are supported by pollen, extinct mammals as glyptodonts, and horses. Arid to semi-arid contexts are supported by some mammalian taxa like *Microcavia, Galea, Lama, Chaetophractus,* and *Lagostomus* among others, that today are extinct in Uruguay but live under such conditions in different regions of South America (Ubilla and Rinderknecht 2014a, b). Pollen from 10 to 11 ka beds indicates a dominance of herbs related to open areas accompanied by lotic contexts suggested by some aquatic weeds. According to the aforementioned numerical ages and the mammal record, the predominant climatic conditions were under the influence of the last phases of the MIS-3 and particularly the MIS-2, corresponding to the last glacial maximum. The predominance of a cold climate can explain the absence of tropical to subtropical mammals in the fossil assemblage.

3.4.4 Other Units

The Raigón Formation has a transitional to fluvial origin (Perea et al. 2013), presumably developed under a humid and seasonally climate (Tófalo et al. 2009) or arid to semi-arid conditions (Bossi et al. 2009; Panario et al. 2014). However, aeolian origin was also proposed for the top of the unit (Bossi et al. 2009). Several fossil vertebrates suggest open areas most likely related to semi-arid environments (Perea and Martínez 2004). If warm and arid conditions were prevalent, and according to the aforementioned OSL ages, correlation with some war episodes like MIS 11, 9 or MIS 7 and 5 is possible.

The consensus on environmental and climatic signals provided by the Libertad Formation is elusive. Semi-arid—including humid episodes—extremely rainy to glacial episodes, alternating episodes of dry and cold to warm and humid, are some of the available suggestions (Panario and Gutiérrez 1999; Bossi et al. 2009; Tófalo et al. 2009; Panario et al. 2014).

References

Abba AM, Poljak S, Superina M (2014) *Chaetophractus villosus*. In: The IUCN red list of threatened species. http://www.iucnredlist.org. Accessed 13 Aug 2015

Antón D (1975) Evolución geomorfológica del norte del Uruguay. Dir Suelos Fert Min Agr Pesca Monevideo 1–28

Austral A (1995) Los cazadores recolectores del sitio estratificado de Paypaso hace 10.000 años. In: Consens M, Lopez J, Curbelo C (eds) Paper presented at the VII Congreso Nacional de Arqueología, Montevideo, 1999

Aznárez J (1945) Apuntes y notas sobre una nomenclatura para los suelos del Uruguay, basada en la geología, con referencia especial a los departamentos de Paysandú, Río Negro, Artigas, Salto y Rivera. Rev Fac Agron 40:67–201

Bossi J, Ferrando L (2001) Carta Geológica del Uruguay. Escala 1.500.000. Geoeditores. Montevideo

Bossi J, Navarro R (1991) Geologia del Uruguay. Dept Publ Univ Rep. 2, Montevideo

Bossi J, Ortiz A, Perea D (2009) Pliocene to middle Pleistocene in Uruguay: a model of climate evolution. Quat Intern 210:37–43

Bradley R (2015) Paleoclimatology: Reconstructing climates of the Quaternary. Elsevier, Amsterdam

Caorsi JH, Goñi JC (1958) Geología Uruguaya. Bol Inst Geol Uruguay 37:1–73

Castiñeira C, Zarate M, Blasi A, Fernicola J, del Puerto L, Inda H, Bracco R, García F (2010) Aportes para una actualización de la correlación entre la Fm. Sopas del Norte de Uruguay – Fm. Luján de la Provincia de Buenos Aires: implicancias arqueológicas. In: Coco G, Feuillet M (eds) Arqueología de Cazadores-Recolectores en la Cuenca del Plata. Centro de Estudios Hispanoamericanos. Ed. Santa Fé

Chimento N, Rey L (2008) Hallazgo de una feca fósil en el Pleistoceno superior-Holoceno inferior del partido de General Guido, provincia de Buenos Aires, Argentina. Rev Mus Arg Cienc Nat 10:239–254

Cid A, Anjos R, Zamboni C, Cardoso R, Muniz M, Corona A, Valladares D, Kovacs L, Macario K, Perea D, Goso Aguilar C, Velazco H (2014) Na, K, Ca, Mg, and U-series in fossil bone and the proposal of a radial diffusion-adsorption model of uranium uptake. J Env Rad 136:131–139

Corona A, Perea D, Toriño P, Goso Aguilar C (2012) Taphonomy, sedimentology and chronology of a fossiliferous outcrop from the continental Pleistocene of Uruguay. Rev Mex Cienc Geol 29:514–525

Corona A, Perea D, MacDonald G (2013) *Catonyx cuvieri* (Xenarthra, Mylodontidae, Scelidotheriinae) from the late Pleistocene of Uruguay, with comments regarding the systematics of the subfamily. J Vert Paleont 33:1214–1225

De Oliveira K, García J, Daners G, Ubilla M, Goso Aguilar C, Bistrichi C (2011) Contribuicao palinoflorística para o cenario do Pleistoceno superior-Holoceno inferior na bacia do Rio Santa Lucia, sul de Uruguai. In: Souza Carvalho I, Kumar N, Strohschoen O, Cunha C (eds) Paleontología: cenários de Vida, vol 3. Interciencia, Rio de Janeiro, pp 175–187

Fariña R, Tambusso S, Varela L, Czerwonogora A, Di Giacomo M, Musso M, Bracco R, Gascué A (2013) Arroyo del Vizcaíno, Uruguay: a fossil-rich 30-ka-old megafaunal locality with cut-marked bones. Proc Royal Soc 281:21132211

Francis J, Mones A (1965) Contribución a la Geología y Paleontología de las Barrancas de San Gregorio, Departamento de San José, República Oriental del Uruguay. Kraglieviana 1(2): 55–85

Frenguelli JJ (1920) Contribución al conocimiento de la geología de Entre Ríos. Bol Acad Nac Cienc Córdoba 24:55–256

Gasparini G, Ubilla M, Tonni EP (2009) Tres especies de tayassuidos (Catagonus wagneri, C. stenocephalus y Tayassu pecari) en el Pleistoceno tardío del norte de Uruguay (Fm. Sopas). In: Paper presented at the Reunión Anual Comunicaciones Asociación Paleontológica Argentina 47, Buenos Aires

Gasparini G, Ubilla M, Tonni EP (2013) The Chacoan peccary, Catagonus wagneri (Mammalia, Tayassuidae) in the late Pleistocene (northern Uruguay, South America): palaeoecological and palaeobiogeographic considerations. Hist Biol 25:679–690

Genise J, Cantil L, Dinghi PA, Sánchez M, Sarzetti L (2013) The aestivation chamber of the giant earthworm Glossoscolex bergi (Glossoscolecidae) in the subtropical rainforest of Misiones (Argentina). Ichnos 20:116–119

Goso H (1972) Cuaternario. Programa de Estudio y Levantamiento de Suelos del Ministerio de Ganadería, Agricultura y Pesca (MGAP), Montevideo, pp 1–12 (unpublished)

Goso H, Bossi J (1966) Cenozoico. In: Bossi J (ed) Geología del Uruguay. Div Publ Universidad de la República, Montevideo, pp 259–301

Goso Aguilar C (2006) Aspectos sedimentológicos y estratigráficos de los depósitos cuaternarios de la costa platense del Departamento de Canelones (Uruguay). Lat Amer J Sed Basin An 13:77–89

Goso Aguilar C, Ubilla M (2004) Los depósitos continentales cuaternarios en el norte de Uruguay: estratigrafía y paleontología. In: Paper presented at the IV Congreso Uruguayo de Geología 13:1–6

Guidón N (1989) Misión de rescate arqueológico Salto Grande. República Oriental del Uruguay, Montevideo

Gutiérrez M, Alberdi MT, Prado JL, Perea D (2005) Late Pleistocene Stegomastodon (Mammalia, Proboscidea) from Uruguay. N Jahrb Geol Palaeontol Mh 11:641–662

Hilbert K (1991) Aspectos de la Arqueología en el Uruguay. Verlang Philipp Von Zabern, Mainz am Rhein

Inda H, del Puerto L (2002) Identificación taxonómica de muestras de material leñoso. Informe inédito. FC. 1–8p

Iriondo M, Kröhling D (2003) A neoformed kaolinitic mineral in the Upper Pleistocene of northeastern Argentina. In: Dominguez E, Mas G, Cravero F (eds) A clay Odyssey. Elsevier, Amsterdam, pp 109–116

Iriondo M, Kröhling D (2008) Cambios ambientales en la cuenca del río Uruguay. Univ Nac Litoral, Santa Fé, Desde dos millones de años hasta el presente

Keuroghlian A, Desbiez A, Reyna-Hurtado R, Altrichter M, Beck H, Taber A, Fragoso J (2013) Tayassu pecari. In: The IUCN Red List of Threatened Species. http://www.iucnredlist.org. Accessed 19 Aug 2015

Klappenbach M, Olazarri J (1986) Notas sobre Strophocheilidae, IV. Eoborus berroi, nueva especie del Mioceno uruguayo. Com Paleont Mus Hist Nat Montevideo I(15):217–225

Lambert R (1940a) Memoria explicativa de un mapa geológico de reconocimiento del Departamento de Paysandú y de los alrededores de Salto. Bol Inst Geol Uruguay 27:1–41

Lambert R (1940b) Memoria explicativa de un mapa geológico de reconocimiento del Departamento de Río Negro. Bol Inst Geol Uruguay 28:1–33

López Romanelli F (2012) El yacimiento "Los Ciervos": hallazgos paleontológicos y arqueológicos en sedimentos de la transición Pleistoceno-Holoceno (Dpto. de Lavalleja, Uruguay). Orígenes 11:1–16

López Mazz J (2013) Early human occupation of Uruguay: radiocarbon database and archaeological implications. Quat Int 301:94–103

López Mazz J, Moreno F, Villamarzo E, Gascue A (2009) Apuntes para una Arqueología costera y del Cabo Polonio. In: López Mass and Gascue A (eds) Arqueología Prehistórica del Uruguay en el Siglo XXI, Biblioteca Nacional, Montevideo, pp 39–65

Martínez S, Rojas A (2004) Quaternary continental mollusks from northern Uruguay: distribution and paleoecology. Quat Int 114:123–128

Martínez S, Ubilla M (2004) El Cuaternario en Uruguay. In: Veroslavsky G, Ubilla M and Martínez S (eds) Cuencas sedimentarias de Uruguay. Geología, Paleontología y Recursos Naturales. Cenozoico, DIRAC-FC, Montevideo, pp 195–227

Martinez S, Veroslavsky G, Verde M (1997) Primer registro del Paleoceno en el Uruguay: paleosuelos calcáreos fosilíferos en la cuenca de Santa Lucía. Re Bras Geoc 27(3):295–302

MEC (1989) Misión de Rescate Arqueológico de Salto Grande. Ministerio de Educación y Cultura, Montevideo

Meneghin H (2004) Artefactos líticos excepcionales del Uruguay. Orígenes 1:1–17

Meneghin H (2006) Un nuevo registro radiocarbónico (C-14) en el yacimiento Urupez II, Maldonado, Uruguay. Orígenes 5:1–7

Meneghin H (2011) Observaciones sobre algunos artefactos líticos discoidales registrados en el Uruguay. Orígenes 10:1–32

Meneghin H (2015) Secuencia cronoestratigráfica de Urupez II. Nuevas dataciones radiométricas. Orígenes 13:1–19

Morosi E, Ubilla M (2014) Preliminary report on isotopic studies (δ ^{13}C) in living and Late Pleistocene ungulates of Uruguay: paleoecological inferences. In: Paper presented at the 4th International Palaeontological Congress, Mendoza, 28 Sep–10 Oct 2015

Naveda A, Thoisy B, Richard-Hansen C, Torres DA, Salas L, Wallance R, Chalukian S, Bustos S (2008) Tapirus terrestris. In: The IUCN Red List of Threatened Species. http://www.iucnredlist.org. Accessed 19 Aug 2015

Panario D, Gutiérrez O (1999) The continental Uruguayan Cenozoic: an overview. Quat Intern 62:75–84

Panario D, Gutiérrez O, Sanchez L, Peel E, Oyhantcabal P, Rabassa J (2014) Ancient Landscapes of Uruguay. In: Rabassa J, Ollier C (eds) Gondwana landscapes in southern South America. Springer Earth System Science, Germany, pp 161–199

Pardiñas U, Ojeda R (2008) Microcavia australis. The IUCN red list of threatened species. In: The IUCN red list of threatened species. http://www.iucnredlist.org. Accessed 19 Aug 2015

Patton JL, Pardiñas U, D´Elía G (2015) Mammals of South America. The University of Chicago Press, New York, Rodents

Perea D (2007) Nothrotherium cf. N. maquinense (Xenarthra, Tardigrada) em la Formacion Sopas (Pleistoceno tardio de Uruguay). Rev Soc Urug Geol 14:5–9

Perea D, Martínez S (2004) Estratigrafía del Mioceno-Pleistoceno en el litoral sur-oeste de Uruguay. In: Veroslavsky G, Ubilla M, Martínez S (eds) Cuencas sedimentarias de Uruguay: geología, paleontología y recursos naturales – Cenozoico. Dirac, Montevideo, pp 105–124

Perea D, Rinderknecht A, Ubilla M, Bostelmann E, Martínez S (2013) Mamíferos y estratigrafía del Neógeno de Uruguay. In: Brandoni J, Noriega I (eds) El Neógeno de la Mesopotamia Argentina, Asociación Paleontológica Argentina Publicación Especial, 14, pp 192–206

Preciozzi F, Spoturno J, Heinzen W, Rossi P (1985) Carta Geológica del Uruguay a escala 1:500.000. Dirección Nacional de Minería y Geología, Montevideo

Prevosti F, Ubilla M, Perea D (2009) Large extinct canids from the Pleistocene of Uruguay: systematic, biogeographic and palaeoecological remarks. Hist Biol 21:79–89

Prosul (2009–2011) Estudo integrado do Cuaternario da Regiao Oeste do Rio Grande do Sul, Regiáo Mesopotamica e pampeana da Argentina e Noroeste do Uruguai. CNPq-Prosul 490299/2008-3, Coordenador: Dr. Ana Ribeiro (FZB)

Quintana C (1996) Diversidad del roedor Microcavia (Caviamorpha, Caviidae) de América del Sur. Mastozoología Neotropical 3:63–86

Rego, N, Perea, D, Toriño, P, Sanchez A (2007) *Chaetophractus villosus* (Desmarest, 1804) (Xenarthra: Dasypodidae) en la Formación Dolores (Pleistoceno Tardío, Uruguay): implicancias paleoambientales. In: Paper presented at the V Congreso Uruguayo de Geología, Facultad de Ciencias, Montevideo, 2–5 August 2007

Rinderknecht A, Blanco E (2008) The largest fossil rodent. Proc Roy Soc Lond 257:923–928

Spoturno J, Oyhantcabal P, Goso Aguilar C, Aubet N, Cazaux S, Huelmo S, Morales E, Loureiro J (2004) Mapa geológico del Departamento de Canelones a escala 1:100.000. CONICYT, Facultad de Ciencias, DINAMIGE, Uruguay

Scherer C (2009) Os Camelidae Lamini (Mammalia, Artiodactyla) do Pleistoceno da America do Sul: aspectos taxonómicos e filogenéticos. Dissertation, UFGRS-IG-PPgraduacao

Suárez R, Lopez J (2003) Archaeology of the Pleistocene-Holocene transition in Uruguay: an overview. Quat Int 109:65–76

Suárez R (2011) Arqueología durante la Transición Pleistoceno Holoceno en Uruguay: Componentes Paleoindios, organización de la tecnología lítica y movilidad de los Primeros Americanos. Archaeopress, British Archaeological Reports (BAR) International Series 2220, Oxford

Suárez R, Santos G (2010) Cazadores-recolectores tempranos, supervivencia de fauna del Pleistoceno (*Equus* y *Glyptodon*), y tecnología lítica durante el Holoceno temprano en la frontera Uruguay-Brasil. Rev Arqueol 23:20–39

Suárez R, Borrero L, Borrazzo K, Ubilla M, Martínez S, Perea D (2014) Archaeological evidences are still missing: Comment on Fariña et al. Arroyo del Vizcaíno Site, Uruguay. Proc Royal Soc B 281:20140449

Tambussi C, Ubilla M, Perea D (1999) The youngest large carnassial bird (Phorusrhacidae, Phorusrhacinae) from South America (Pliocene-Early Pleistocene, Uruguay). J Vert Paleont 19:404–406

Tambussi C, Ubilla M, Acosta Hospitaleche C, Perea D (2005) Fossil records and palaeoenvironmental implications of *Chloephaga picta* (Gmelin, 1789) (Magellan Goose) and *Cariama cristata* (Linnaeus, 1766) (Seriema) from the Late Pleistocene of Uruguay. N J Geol Palaeontol Mh 5:257–268

Tambussi C, Acosta Hospitaleche C, Rinderknecht A, Ubilla M (2009) Parrots (Aves, Psittaciformes) in the Pleistocene of Uruguay. Ameghiniana 46:431–435

Tófalo OR, Orgeira M, Morrás H, Vázquez C, Sánchez L, Pecoits E, Aubet N, Sánchez G, Zech W, Moretti L (2009) Geological, pedological and paleomagnetic study of the late Cenozoic sedimentary sequence in southwestern Uruguay, South America. Quat Intern 210: 6–17

Ubilla M (1999) Dataciones radiocarbónicas (C14) para la Fm. Dolores (Río Santa Lucía, Dpto. de Canelones y A° Gutiérrez Chico, Dpto. de Río Negro) y comentarios sobre la fauna de vertebrados asociada. Rev Soc Geol Uruguay 6:48–53

Ubilla M (2001) Comment on "The continental Uruguayan cenozoic: an overview" by D. Panario & O. Gutiérrez (Quaternary International 62, 75–84). Quat Intern 76–77:59–260

Ubilla M (2004) Mammalian biostratigraphy of Pleistocene fluvial deposits in northern Uruguay, South America. Proc Geol Assoc Lond 115:347–357

Ubilla M (2008) Postcranial morphology of the extinct caviine rodent *Microcavia criolloensis* (late Pleistocene, South America). Zool J Linn Soc 154(4):795–806

Ubilla M, Perea D (1999) Quaternary vertebrates of Uruguay: biostratigraphic, biogeographic and climatic overview. Quat South Amer Antarct Pen 12:75–90

Ubilla M, Piñeiro G, Quintana C (1999) A new extinct species of the genus *Microcavia* (Rodentia, Caviidae) from the Upper Pleistocene of the northern basin of Uruguay (South America) with paleobiogeographic and paleoenvironmental comments. Studies Neotrop Fauna Environ 34:141–149

Ubilla M, Perea D, Goso Aguilar C, Lorenzo N (2004) Late Pleistocene vertebrates from northern Uruguay: tools for biostratigraphic, climatic and environmental reconstruction. Quat Intern 114:129–142

Ubilla M, Perea D, Rinderknecht A, Corona A (2009) Pleistocene mammals from Uruguay: biostratigraphic, biogeographic and environmental connotations. In: Ribeiro A, Girardi S, Saldanha C (eds) Quaternario de Rio Grande do Sul. Integrando Conhecimentos Monografías Sociedade Brasileira de Paleontología, Porto Alegre, pp 217–230

Ubilla M, Perea D, Bond M, Rinderknecht A (2011) The first cranial remains of the Pleistocene Proterotheriid *Neolicaphrium Frenguelli*, 1921 (Mammalia, Litopterna): a comparative approach. J Vert Paleontol 31(1):193–201

Ubilla M, Rinderknecht A, Corona A, Perea D (2013) Late records of extinct mammals, environments and climate from Late Pleistocene-Early Holocene fluvial beds (Southern Uruguay, South America). In: Paper presented at the 61th Symposium on Vertebrate Palaeontology and Comparative Anatomy and 22th Symposium of Palaeontological Preparation and Conservation Annual Meeting, Edinburgh

Ubilla M, Rinderknecht A (2014a) Comparative analysis of Galea (Rodentia, Caviidae) and expanded diagnosis of *Galea ortodonta* Ubilla & Rinderknecht, 2001 (Late Pleistocene, Uruguay). Geobios 47:255–269

Ubilla M, Rinderknecht A (2014b) The plains vizcacha *Lagostomus maximus* (Desmarest) in the Late Pleistocene of Uruguay: biogeographic and environmental considerations. In: Paper presented at the 4th International Palaeontological Congress, Mendoza, 28 Sep–10 Oct 2015

Ubilla M, Corona A, Rinderknecht A, Perea D, Verde M (accepted 2015) MIS-3 and continental beds from northern Uruguay (Sopas Formation): palaeontology, chronology and climate. In: Gasparini GM, Rabassa J, Deschamps MC, Tonni EP (eds) Marine Isotope Stage 3 in Southern South America 60 KA B.P.-30 KA B.P. Editorial Springer International Publishing AG, Germany

Verde M, Ubilla M (2002) Mammalian carnivore coprolites from the Sopas Formation (Upper Pleistocene, Lujanian Stage), Uruguay. Ichnos 9:77–80

Verde M, Ubilla M, Jiménez J, Genise J (2007) A new earthworm trace fossil from palaeosols: aestivation chambers from the late pleistocene Sopas formation of Uruguay. Palaeogr Palaeoclimatol Palaeoecol 243:339–347

Veroslavsky G, Montaño J (2004) Sedimentología y Estratigrafía de la Formación Salto (Pleistoceno). In: Veroslavsky G, Ubilla M, Martínez S (eds) Cuencas sedimentarias de Uruguay: geología, paleontología y recursos naturales – Cenozoico. Dirac, Montevideo, pp 147–166

Veroslavsky G, Ubilla M (2007) A "snapshot" of the evolution of the Uruguay River (del Plata Basin): the Salto depositional sequence (Pleistocene, Uruguay, South America). Quat Sci Rev 26:2913–2923

Walther K (1931) Sedimentos gelíticos y clastogelíticos del Cretácico superior Terciario uruguayos. Observaciones referentes a algunos productos de desintegración moderna del País. Bol Inst Geol Perforaciones 13:1–142

Chapter 4
The Quaternary of Uruguay at Regional Scale

Abstract The marine and estuarine beds of Uruguay may be roughly ascribed to two groups, one belonging to the Late Pleistocene and other to the Holocene. In the past, many names and units were introduced in the literature, mainly designating facies, but currently, and thanks to a good data base of numerical ages, we can simplify this former, almost anarchic situation, with the correlation by numerical ages made much easier. In Uruguay, the Pleistocene deposits correspond to MIS 5e, and the Holocene ones to its middle and late part, with equivalent deposits in southern Brazil and northern Argentina. The identification of units in these areas suffered the same vices; consequently, it is wise to correlate the quaternary deposits on an age-only based framework. The continental beds correlate with the Pampean and Mesopotamian areas of Argentina and southern Brazil. The fluvial Salto Formation of western Uruguay has been chronologically/lithologically correlated with beds of the Argentinian Mesopotamia (El Palmar Formation, Salto Chico Formation and Ituzaingó Formation). The sedimentary beds of northern Uruguay (Sopas Formation) are lithologically and chronologically correlated with the Touro Passo Formation of southern Brazil, reflecting fluvial and aeolian depositional processes, mostly related to the MIS-3. The sedimentary beds of southern Uruguay, the Dolores Formation, correlates in age with Late Pleistocene beds of the Pampean region of Argentina (the Guerrero Member of the Luján Formation), mostly related to the influence of the MIS-2.

Keywords Lithological correlation · Chronological correlation · Continental pleistocene deposits · Marine pleistocene deposits · Marine holocene beds

4.1 Correlations: Problems and Limitations

4.1.1 Marine Beds

As stated in Sect. 2.1 and summarized in Fig. 1.2, a large number of lithostratigraphic or lithostratigraphic-like units have been proposed through the relative short

© United Nations University 2016
M. Ubilla and S. Martínez, *Geology and Paleontology of the Quaternary of Uruguay*, SpringerBriefs in Earth System Sciences,
DOI 10.1007/978-3-319-29303-5_4

history of geological research in Uruguay. Worse, several of these units were established in studies considered "gray literature" and with local points of view, and seeing a new "ingression" in every outcrop or facies change. This situation generated great confusion until a goodly number of radiocarbon ages were obtained, showing the existence of two gross episodes: one Late Pleistocene (Chuy Formation) and other Middle-Late Holocene (Villa Soriano Formation). Goñi and Hoffstetter (1964) and Martínez et al. (2001, 2006) give accounts of the old nomenclature.

Regional correlation was simultaneously easy as well as difficult, for the same reasons that affected the local one: until having a reasonable set of radiocarbon ages, an intricate combination of names were used to correlate units that were, finally, sedimentary facies taken on a small scale. For example, Ameghino (1889) coined the term "*Belgranense*" for a Pleistocene transgression differentiated from a Holocene one in Argentina. However, subsequently, local names and small lithological units appeared, and it became very difficult to correlate the units across the Rio de la Plata (see names in Aguirre and Whatley 1995; Isla et al. 2000; Aguirre and Fucks 2004; Martínez and Ubilla 2004). Lately, the simplification was given by numerical ages, again to correlate with the latest Pleistocene transgression in the region that corresponds to MIS5e. There was a similar situation with the Holocene deposits, but now we have a simpler but more reliable correlation with the Argentinean deposits. There had also been names that referred to local facies, "*platense*" for pelitic deposits, and "*querandinense*" for sandy ones; in fact, both are present in the Pleistocene and Holocene of Uruguay and Argentina.

Although the marine and estuarine Quaternary is widely represented along the coast of Brazil, it is reasonable that the correlations have been made with its southern portion, namely the state of Rio Grande do Sul. There, Pleistocene and Holocene sedimentites are found, too, but their systematization has traditionally been done from a geomorphological perspective. In fact, the eastern Uruguayan coast is the extreme of the Rio Grande do Sul coastal plain. The Patos Group safely correlates with both the Chuy and Villa Soriano Formations (Forti-Esteves 1974; Martínez 1990). Recently, Lopes et al. (2013, 2014) claimed the finding of deposits belonging to MIS-7 very near the border with Uruguay, mainly based on ESR dating on shells, but also with stratigraphic arguments. Oddly, there is still no evidence of it across the border.

4.1.2   Continental Beds

In the available literature, proposals about correlation are often imprecise, because on occasion it is not clear whether authors are talking about lithological or chronological correlation—or both.

The fluvial Salto Formation of western Uruguay is usually lithologically correlated with a fluvial and transitional unit of southern Uruguay, the Raigón Formation (Fig. 1.2) (Goso 1965; Goso and Bossi 1966; Panario and Gutiérrez

1999; Martínez and Ubilla 2004). However, the most complex topic is the regional correlation of this unit, because the consensus on nomenclature and chronology of the sedimentary units involved in the Argentinian Mesopotamian area is often elusive. The Salto Chico Formation (Rimoldi 1962) and the El Palmar Formation (Iriondo 1980) from the Argentinean Mesopotamia have lithological features that are similar to those of the Salto Formation, and can be lithologically correlated with certainty (Veroslavsky and Montaño 2004; Iriondo and Kröhling 2008). However, it is possible that these units could have different ages (Iriondo 1999; Iriondo and Santi 2000). On the other hand, the Ituzaingó Formation was correlated or identified with the Salto Chico Formation (Gentili and Rimoldi 1979; Herbst 2000), which was in turn also correlated with the Salto Chico and Salto Formations (Panario et al. 2014). Noticeably, there are differing opinions on the age of the Ituzaingó Formation (Late Miocene to Pleistocene) (Cione et al. 2000; Herbst 2000). In southern Brazil, there are geologically equivalent sedimentary beds that yield similar paleontological content with the Salto Formation (Gentili et al. 1974; Bolzon and Oliveira 2000; Iriondo and Kröhling 2008). Aside from this complicated situation, there is a consensus that these units likely represent Pleistocene evolutionary phases of the Uruguay and Paraná fluvial system, which are today members of the Cuenca del Río de la Plata (Veroslavsky and Ubilla 2007).

Bombín (1975, 1976) proposed a lithological correlation between the Sopas Formation of northern Uruguay and the Touro Passo Formation in southern Brazil. For the Brazilian unit, numerical ages indicate a similar Late Pleistocene age (Kerber et al. 2011); it yields not only similar lithological features, but also shares some mammalian taxa with the Sopas Formation (Kerber and Oliveira 2008). According to numerical ages (Miño-Boilini et al. 2006 and references therein), a likely chronological correlation of the Sopas Formation with units that outcrop in northern Argentina (the Toropí and Yupoí Formations sensu Herbst and Alvarez 1975) should also be considered. The Sopas Formation correlates in time with the "*Secuencia deposicional Luján Verde Inferior*" and in part with the "*Secuencia deposicional Luján Rojo*" from the Argentinian Pampean region (Toledo 2011). It is possible to postulate a partial chronological correlation with the La Chumbiada Member of the Luján Formation of the Pampean region, dated ca. 30 ka, with some mammals related to temperate climate (Tonni et al. 1999; Ubilla 2004).

The Dolores Formation in southern Uruguay partially correlates in age with the Guerrero Member of the Luján Formation of the Buenos Aires Province (Ubilla et al. 2009). The beds of this unit are fluvial deposits with ages ranging from 21 to 10 ka and yield a vertebrate assemblage reflecting various climatic episodes, in particular those related to the Last Glacial Maximum (Iriondo and García 1993; Tonni et al. 1999; Tonni 2009; Tonni et al. 2003). It is also possible to consider a correlation with the "*Secuencia deposicional Luján rojo*," "*Secuencia deposicional Luján verde superior*," and the basal portion of the "*Secuencia La Plata*," with ages ranging from 30 to 12 ka (Toledo 2011). The aforementioned sedimentological facies of northern Uruguay, with ages ranging from 12 to 8 ka (see Chap. 3), can also be partially chronologically correlated with the Dolores beds.

References

Aguirre M, Fucks E (2004) Moluscos y Paleoambientes del Cuaternario Marino en el sur de Entre
    Ríos y Litoral Bonaerense. In: Aceñolaza F (ed) Temas de la Biodiversidad del Litoral Fluvial
    Argentino, vol 12. INSUGEO, Miscelánea, Entre Ríos, pp 55–70
Aguirre ML, Whatley RC (1995) Late Quaternary marginal marine deposits and palaeonvironments
    from northeastern Buenos Aires Province, Argentina: a review. Quat Sci Revi 14:223–254
Ameghino F (1889) Contribución al conocimiento de los mamíferos fósiles de la República
    Argentina. Actas Acad Na Cienc Córdoba 6:1–1027
Bolzon RT, Oliveira E (2000) Associação de lenhosfósseis da região de Uruguaiana, estado do Rio
    Grande do Sul, Brasil: considerações sobre geología e tafonomía. Geocienc 5:160–163
Bombín M (1975) Afinidade paleoecológica, cronológica e estratigráfica do componente de
    megamamíferos na biota do quaternario terminal da Província de Buenos Aires, Uruguai e Rio
    Grande do Sul (Brasil). Com Mus Cienc PUCRGS, Porto Alegre 9:1–28
Bombín M (1976) Modelo paleoecológico evolutivo para o NeoCuaternario da regiâo da
    campanha oeste do Rio Grande do Sul (Brasil). A FormaçaoTouroPasso, seu conteúdo
    fossilífero e a pedogénesis posdeposicional. Com Mus Cienc PUCRGS, Porto Alegre 15:1–90
Cione A, Azpelicueta M, Bond M, Carlini A (2000) Miocene vertebrates from Entre Ríos
    province, eastern Argentina. In: Herbst R (ed) Aceñolaza F. El Neógeno Argentino INSUGEO,
    Tucumán, pp 191–237
Forti-Esteves IRS (1974) Bioestratigrafia e Paleoecologia (Mollusca) do Quaternário da Planicie
    costeira do Rio Grande do Sul (Brasil). Anais 28º Congresso Brasileiro de Geologia, Porto
    Alegre, 133–149
Gentili CA, Rimoldi H (1979) Mesopotamia. Paper presented at the 2° Simposio Geología
    Regional Argentina, Córdoba
Gentili CA, Rosenmann ML, Lourenco AM (1974) Características geológicas generales del
    territorio brasileño en la faja adyacente al curso medio del Río Uruguay. Rev Asoc Geol
    Argentina 24(2):223–230
Goñi JC, Hoffstetter R (1964) Uruguay. Lexique Stratigraphique. International, 5 Amérique
    Latine, 9a. CNRS, Paris
Goso H (1965) El Cenozoico en el Uruguay. Inst Geol Uruguay 1–36
Goso H, Bossi J (1966) Cenozoico. In: Bossi J (ed) Geología del Uruguay. Div Publ Universidad
    de la República, Montevideo, pp 259–301
Herbst R (2000) La Formación Ituzaingó (Plioceno). Estratigrafía y distribución. In: Aceñolaza F,
    Herbst R (eds) El Neógeno argentina, INSUGEO, Tucumán, pp 181–190
Herbst R, Alvarez B (1975) Nota sobre dos formaciones del Cuaternario en Corrientes. An Acad
    Bras Cienc 47:33–37
Iriondo M (1980) El Cuaternario de Entre Ríos. Rev Asoc Cienc Nat Litoral 11:125–141
Iriondo M (1999) El Cuaternario del Chaco y Litoral. In: Caminos R (ed): Geología Argentina
Iriondo M, García N (1993) Climatic variations in the Argentine plains during the last 18,000 year.
    Palaeogeogr Palaeoclim Palaeoecol 101:209–220
Iriondo M, Kröhling D (2008) Cambios ambientales en la cuenca del Río Uruguay. Desde dos
    millones de años hasta el presente. Univ Nac Litoral, Santa Fé
Iriondo M, Santi M (2000) La Formación Salto Chico en el subsuelo de Entre Ríos. Paper
    presented at the II Congreso Latinoamericano de Sedimentología
Isla F, Rutter N, Schnack EJ, Zarate MA (2000) La transgresión Belgranense en Buenos Aires.
    Una revisión a cien años de su definición. Rev Cuat Cienc Amb 1:3–14
Kerber L, Oliveira E (2008) Fósseis de vertebrados da Formaçao Touro Passo (Pleistoceno superior),
    Rio Grande do Sul, Brazil: atualizaçaõ dos dados e novas contribucões. GAEA 4:49–64
Kerber L, Kinoshita A, José F, Graciano A, Oliveira E, Baffa O (2011) Electron Spin Resonance
    Dating of the Southern Brazilian Pleistocene mammals from Touro Passo Formation, and
    remarks on the geochronology, fauna and paleoenvironments. Quat Int 245:201–208

Lopes RP, Simone LRL, Dillenburg SR, Schultz CL, Pereira JC (2013) A Middle Pleistocene marine molluscan assemblage from the southern coastal plain of Rio Grande do Sul state, Brazil. Rev Bras Paleontol 16:343–360

Lopes RP, Kinoshita A, Baffa O, Figueiredo AMG, Dillenburg SR, Schultz CL, Pereira JC (2014) ESR dating of Pleistocene mammals and marine shells from the coastal plain of Rio Grande do Sul state, southern Brazil. Quat Int 352:124–134

Martínez S (1990) Taphonomy and Paleoecology of Holocene mollusks from the western margin of the Merin Lagoon (Villa Soriano Fm., Uruguay). Quat South Am Antarctic Pen 7:121–135

Martínez S, Ubilla M (2004) El Cuaternario de Uruguay. In: Veroslavsky G, Ubilla M, Martínez S (eds) Cuencas sedimentarias de Uruguay. Cenozoico, Dirac, Montevideo, pp 195–227

Martínez S, Ubilla M, Verde M, Perea D, Rojas A, Guérèquiz R, Piñeiro G (2001) Paleoecology and geochronology of Uruguayan coastal marine Pleistocene deposits. Quat Res 55:246–254

Martínez S, Rojas A, Ubilla M, Verde M, Perea D, Piñeiro G (2006) Molluscan assemblages from the marine Holocene of Uruguay: composition, geochronology and paleoenvironmental signals. Ameghiniana 43:385–398

Miño-Boilini A, Cerdeño E, Bond M (2006) Revisión del género *Toxodon* Owen 1837 (Notoungulata, Toxodontidae) en el Pleistoceno de las provincias de Corrientes, Chaco y Santa Fé, Argentina. Rev Española Paleont 21:93–103

Panario D, Gutiérrez O (1999) The continental Uruguayan Cenozoic: an overview. Quat Intern 62:75–84

Panario D, Gutiérrez O, Sanchez L, Peel E, Oyhantcabal P, Rabassa J (2014) Ancient Landscapes of Uruguay. In: Rabassa J, Ollier C (eds) Gondwana Landscapes in southern South America, Springer Earth System Science, Germany, pp 161–199

Rimoldi HV (1962) Aprovechamiento del río Uruguay en la zona de Salto Grande: estudio geotectônico-geológico para la presa de compensación proyectada en Paso Hervidero (província de Entre Ríos). Paper presented at the Primeras Jornadas de Geología Argentina, Buenos Aires, 1962

Toledo M (2011) El legado lujanense de Ameghino: revisión estratigráfica de los depósitos pleistocenos-holocenos del Valle del Río Luján en su sección tipo. Registro paleoclimático en la pampa de los estadios OIS 4 al OIS 1. RevAsocGeol Argentina 68:121–167

Tonni EP (2009) Los mamíferos del cuaternario de la región pampeana de Buenos Aires, Argentina. In: Ribeiro A, Girardi S, Saldanha C (eds) Quaternario de Rio Grande do Sul. Integrando Conhecimentos, Porto Alegre, pp 193–205

Tonni EP, Cione A, Figini A (1999) Predominance of arid climates indicated by mammals in the pampas of Argentina during the late Pleistocene and Holocene. Palaeogeogr Palaeoclim Palaeoecol 147:257–281

Tonni EP, Huarte R, Carbonari J, Figini A (2003) New radiocarbon chronology for the Guerrero Membero of the Luján Formation (Buenos Aires, Argentina): palaeoclimatic significance. Quat Intern 109–110:45–48

Ubilla M (2004) Mammalian biostratigraphy of Pleistocene fluvial deposits in northern Uruguay, South America. Proc Geologists′ Assoc London 115:347–357

Ubilla M, Perea D, Rinderknecht A, Corona A (2009) Pleistocene mammals from Uruguay: biostratigraphic, biogeographic and environmental connotations. In: Ribeiro A, Girardi S, Saldanha C (eds) Quaternario de Rio Grande do Sul. Integrando Conhecimentos Monografías Sociedade Brasileira de Paleontología, Porto Alegre, pp 217–230

Veroslavsky G, Montaño J (2004) Sedimentología y Estratigrafía de la Formación Salto (Pleistoceno). In: Veroslavsky G, Ubilla M, Martínez S (eds) Cuencas sedimentarias de Uruguay: geología, paleontología y recursos naturales—Cenozoico. Dirac, Montevideo, pp 147–166

Veroslavsky G, Ubilla M (2007) A "snapshot" of the evolution of the Uruguay River (del Plata Basin): the Salto depositional sequence (Pleistocene, Uruguay, South America). Quat Sci Rev 26:2913–2923

Appendix 1
Mollusks of the Chuy Formation

Mollusks of the Chuy Formation (Martínez et al. 2001, 2013; Rojas and Urteaga 2011)

	NP	LC
Tegula patagonica	0	1
Heleobia australis	1	0
Bostrycapulus odites	0	1
Finella dubia	0	1
Bittiolum varium	0	0
Seila adamsii	0	0
Iselica globosa	0	1
Chrysallida gemmulosa	0	1
Boonea jadisi	0	1
Turbonilla uruguayensis	0	1
Turbonilla americana	0	1
Siphonaria lessoni	1	0
Olivella sp.	0	1
Stramonita haemastoma	1	0
Costoanachis sertulariarum	0	1
Parvanachis isabellei	0	1
Buccinanops globulosus	1	0
Cylichnella bidentata	1	0
Lunarca ovalis	1	1
Noetia bisulcata	1	0
Limaria sp.	0	1
Ostreola equestris	1	1
Plicatula gibbosa	0	1
Aequipecten tehuelchus	0	1
Trachycardium muricatum	0	1
Laevicardium sp.	0	1
Brachidontes sp.	1	0

(continued)

© United Nations University 2016
M. Ubilla and S. Martínez, *Geology and Paleontology of the Quaternary of Uruguay*, SpringerBriefs in Earth System Sciences,
DOI 10.1007/978-3-319-29303-5

(continued)

	NP	LC
Mactra isabelleana	1	1
Semele proficua	0	1
Abra lioica	0	1
Tagelus plebeius	1	0
Chione subrostrata	1	0
"Clausinella gayi"	0	1
Anomalocardia brasiliana	1	0
Pitar rostratus	1	0
Caryocorbula caribaea	1	1
Chaetopleura isabellei	0	1
Chaetopleura angulat	0	1
Chaetopleura asperrima	0	1
Ischnochiton striolatus	0	1

NP Nueva Palmira; *LC* La Coronilla

Appendix 2
Mollusks of the Villa Soriano Formation

Mollusks of the marine Holocene of Uruguay known at the species level (Martínez et al. 2013)

Lottia subrugosa
Diodora patagonica
Calliostoma jucundum
Photinula blakei
Halistylus columna
Tegula patagonica
Echinolittorina lineolata
Heleobia australis
Heleobia charruana
Bostrycapulus odites
Crepidula protea
Notocochlis isabelleana
Cabestana felipponei
Cymatium parthenopeum
Epitonium albidum
Epitonium georgettinum
Cerithiopsis greenii
Seila adamsii
Triphora medinae
Marshallora nigrocinta
Iselica globosa
Boonea jadisi
Boonea seminuda
Turbonilla abrupta
Turbonilla dispar
Turbonilla uruguayensis
Parodizia uruguayensis
Siphonaria lessoni
Risomurex nicocheanus
Urosalpinx haneti

© United Nations University 2016
M. Ubilla and S. Martínez, *Geology and Paleontology of the Quaternary of Uruguay*, SpringerBriefs in Earth System Sciences, DOI 10.1007/978-3-319-29303-5

Urosalpinx cala
Stramonita haemastoma
Costoanachis sertulariarum
Parvanachis isabellei
Parvanachis obesa
Buccinanops cochlidium
Buccinanops deformis
Buccinanops duartei
Buccinanops globulosus
Buccinanops monilifer
Buccinanops uruguayensis
Olivella puelcha
Olivella tehuelcha
Olivancillaria carcellesi
Olivancillaria deshayesiana
Olivancillaria urceus
Olivancillaria uretai
Olivancillaria auricularia
Zidona dufresnei
Pachycymbiola brasiliana
Prunum martini
Conus platensis
Pleurotomella aguayoi
Spirotropis patagonica
Terebra gemmulata
Bulla occidentalis
Cylichnella bidentata
Acteocina candei
Ennucula puelcha
Adrana electa
Malletia cumingii
Scapharca brasiliana
Lunarca ovalis
Noetia bisulcata
Glycymeris longior
Mytilus edulis
Brachidontes rodriguezii
Brachidontes darwinianus
Lithophaga patagonica
Ostrea puelchana
Ostreola equestris
Crassostrea rhizophorae
Plicatula gibbosa

Aequipecten tehuelchus
Zygochlamys patagonica
Carditamera plata
Pleuromeris sanmartini
Crassinella maldonadoensis
Felaniella vilardeboena
Diplodonta patagonica
Phlyctiderma semiasperum
Trachycardium muricatum
Mactra isabelleana
Mactra marplatensis
Mactra janeiroensis
Mactra guidoi
Raeta plicatella
Mesodesma mactroides
Solen tehuelchus
Donax hilairea
Tagelus plebeius
Semele proficua
Abra lioica
Abra uruguayensis
Macoma uruguayensis
Tellina petitiana
Angulus gibber
Strigilla carnaria
Petricola lapicida
Petricola dactylus
Petricolaria stellae
Chionecancellata
Anomalocardia brasiliana
Tivela zonaria
Tivela isabelleana
Pitar rostratus
Amiantis purpurata
Retrotapes exalbidus
Ameghinomya antiqua
Caryocorbula caribaea
Corbula lyoni
Corbula patagonica
Corbula tryoni
Erodona mactroides
Sphenia fragilis
Barnea lamellosa

Pholas campechiensis
Periploma compressum
Thracia similis
Bushia rushii
Conus semicoronatus

Appendix 3
Non-molluskan Macrofossil Taxa of the Villa Soriano

Selected non-molluskan macrofossil taxa of the marine Holocene of Uruguay. (Lorenzo and Verde 2004; Martínez 1983; Martínez and Mooi 2005; Verde 1994, 2003; Figueiras 1962; Figueiras and Broggi 1967; Verde et al. 1998; Perea and Ubilla 1981; Piñeiro and Verde 1999; Rinderknecht et al. 2005; Mones 1993)

Cnidaria
Astrangia brasiliensis

Annelida
Serpula sp.

Bryozoa
Membranipora tehuelcha

Crustacea
Balanomorpha
Balanus sp.
Decapoda
Chasmagnathus granulata
Munidopsis sp.
Platyxanthus sp.

Echinodermata
Mellitidae
Encope sp.
Mellita sp.

Chondrichthyes
Carcharias sp., *Carcharhinus* sp., *Myliobatis* sp.

Osteichthyes
Micropogonias furnieri, Pogonias cromis, Paralichthys sp., *Mugil* sp.

Aves
Rhea sp.

M. Ubilla and S. Martínez, *Geology and Paleontology of the Quaternary of Uruguay*, SpringerBriefs in Earth System Sciences, DOI 10.1007/978-3-319-29303-5

Ichnofossils

Soft ground:
Skolithos, Thalassinoides, Ophiomorpha, Conichnus, "Tagelus traces,"
Cylindricum, Rosselia? Ex *Cylindrichnus, Helicodromites, Palaeophycus,
Planolites*

Hard ground:
Entobia, Maeandropolydora, Oichnus, Radulichnus, and briozooan holes.

References

Figueiras A (1962) Sobre nuevos hallazgos de moluscos subfósiles de la Transgresión Querandina. Com Soc Malac Uruguay 1:53–68

Figueiras A, Broggi J (1967) Estado actual de nuestros conocimientos sobre los moluscos fósiles del Uruguay Parte I. Com Soc Malac Uruguay 2:147–187

Lorenzo N, Verde M (2004) Estructuras de bioerosión en moluscos marinos de la Formación Villa Soriano (Pleistoceno tardío—Holoceno) de Uruguay. Rev Bras Paleont 7:319–328

Martínez S (1983) Invertebrados fósiles holocenos del Departamento de Rocha, Uruguay (Mollusca y Crustacea). Res Com Jorn Cienc Nat 3:46–47

Martínez S, Mooi R (2005) Extinct and extant sand dollars (Clypeasteroida: Echinoidea) from Uruguay. Rev Biol Trop/Int J Trop Biol 53 Suppl 3:1–7

Martínez S, Ubilla M, Verde M, Perea D, Rojas A, Guérèquiz R, Piñeiro G (2001) Paleoecology and geochronology of Uruguayan coastal marine Pleistocene deposits. Quat Res 55:246–254

Martínez S, del Río CJ, Rojas A (2013) Biogeography of the quaternary molluscs of the Southwestern Atlantic ocean. SpringerBriefs in earth system sciences. springer, Heidelberg

Mones A (1993) Notas paleontológicas uruguayas, V. Primer hallazgo de *Rhea* sp. fósil en el Uruguay (Aves: Struthioniformes: Rheidae). Com Paleont Mus Hist Nat 2:65–68

Perea D, Ubilla M (1981) Estudio preliminar de la ictiofauna fósil marina de las costas del Dpto. de Rocha, Uruguay. Res Com Jorn Cienc Nat Uruguay 2:25–26

Piñeiro G, Verde M (1999) Osteología y paleoecología de Sciaenidos (Teleostei, Perciformes) de la Formación Villa Soriano (Holoceno) de Uruguay. Rev Univ Guarulhos Geociências 4 (6):19–27

Rinderknecht A, Piñeiro G, Verde M, Rojas A (2005) New fish remains from marine deposits of Uruguay. Paper presented at the VIII Jornadas de Zoología del Uruguay, Montevideo

Rojas A, Urteaga D (2011) Late Pleistocene and Holocene chitons (Mollusca, Polyplacophora) from Uruguay: palaeobiogeography and palaeoenvironmental reconstruction in mid latitudes of the Southwestern Atlantic. Geobios 44:377–386

Verde M (1994) Icnofósiles del Holoceno marino de las regiones Sur y Suroeste de Uruguay, Formación Villa Soriano. Paper presented at the VI Congreso Argentino de Paleontología y Bioestratigrafía, Trelew, Chubut

Verde M (2003) Marine invertebrate ichnofossils from the Quaternary of Uruguay. Paper presented at the VII Jornadas de Zoología del Uruguay

Verde M, Delgado E, Piñeiro G (1998) *Chasmagnathus granulata* Dana, 1851 (Decapoda, Grapsidae) en la Formación Villa Soriano (Holoceno) del Uruguay. Paper presented at the II Congreso Uruguayo de Geología, Montevideo

Index

© United Nations University 2016
M. Ubilla and S. Martínez, *Geology and Paleontology of the Quaternary of Uruguay*, SpringerBriefs in Earth System Sciences,
DOI 10.1007/978-3-319-29303-5